For EU product safety concerns:
IN TECH d.o.o., Prolaz Marije Krucifikse Kozulić 3, 51000 Rijeka, Croatia,
info@intechopen.com or visit our website at intechopen.com.

Meet the editor

Dr. Daniel Glossman-Mitnik is now a Titular Researcher at the Centro de Investigación en Materiales Avanzados (CIMAV), Chihuahua, Mexico as well as a National Researcher of Level III of the Consejo Nacional de Ciencia y Tecnología, Mexico. His research interest focuses on Computational Chemistry and Molecular Modeling of molecular systems of pharmacological, food, and alternative energy interests by resorting to DFT and Conceptual DFT. He has authored and coauthored more than 200 peer-reviewed papers and 10 book chapters. He has delivered speeches at many international and domestic conferences. He serves as the referee for more than 60 international journals, books, and research proposals.

Contents

Preface

Density Functional Theory (or DFT for short) is a potent methodology useful for calculating and understanding the molecular and electronic structure of atoms, molecules, clusters, and solids. Its use relies not only in the ability to calculate the molecular properties of the species of interest but also provides interesting concepts that allow a better comprehension of the chemical reactivity of the studied systems.

This book represents an attempt to present examples on the utility of DFT for the understanding of the chemical reactivity through descriptors that constitute the basis of the so called Conceptual DFT (sometimes also named as Chemical Reactivity Theory). These descriptors provide a qualitative and quantitative view of the problem and an updated review of this is presented in the first chapter within the Concepts section.

The Applications section contains chapters showing the application of the theory and its related computational procedures in the determination of the molecular properties of different systems of academic and industrial interest.

I would like to express my sincere gratitude to all authors who contributed to this book: Paul Geerlings, Stijn Fias, Thijs Stuyver, Paul Ayers, Robert Balawender, Frank De Proft, Alejandro E. Pérez, Rafael Ribadeneira, Siwarut Siwaipram, Sarawoot Impeng, Philippe A. Bopp, Sareeya Bureekaew, Manuel Antuch, Pierre Millet, Francisco Colmenero, Sergio Ricardo De Lázaro and Ruby Srivastava.

Finally, my warmest thanks go to my beloved wife Carmen and to the memories of my late parents, Sofía and Miguel. I am also grateful for the financial support from the Consejo Nacional de Ciencia y Tecnología (CONACYT), Mexico.

Dr. Daniel Glossman-Mitnik
Laboratorio Virtual NANOCOSMOS
Departamento de Medio Ambiente y Energía
Centro de Investigación en Materiales Avanzados
Chihuahua, Mexico

Section 1

Concepts

Chapter 1

New Insights and Horizons from the Linear Response Function in Conceptual DFT

Paul Geerlings, Stijn Fias, Thijs Stuyver, Paul Ayers,
Robert Balawender and Frank De Proft

Additional information is available at the end of the chapter

http://dx.doi.org/10.5772/intechopen.80280

Abstract

An overview is given of our recent work on the linear response function (LRF) $\chi(\mathbf{r}, \mathbf{r}')$ and its congener, the softness kernel $s(\mathbf{r}, \mathbf{r}')$, the second functional derivatives of the energy E and the grand potential Ω with respect to the external potential at constant N and μ, respectively. In a first section on new insights into the LRF in the context of conceptual DFT, the mathematical and physical properties of these kernels are scrutinized through the concavity of the $E = E[N, v]$ and $\Omega = \Omega[\mu, v]$ functionals in $v(\mathbf{r})$ resulting, for example, in the negative semidefiniteness of χ. As an example of the analogy between the CDFT functionals and thermodynamic state functions, the analogy between the stability conditions of the macroscopic Gibbs free energy function and the concavity conditions for Ω is established, yielding a relationship between the global and local softness and the softness kernel. The role of LRF and especially the softness kernel in Kohn's nearsightedness of electronic matter (NEM) principle is highlighted. The first numerical results on the softness kernel for molecules are reported and scrutinized for their nearsightedness, reconciling the physicists' NEM view and the chemists' transferability paradigm. The extension of LRF in the context of spin polarized conceptual DFT is presented. Finally, two sections are devoted to 'new horizons' for the LRF. The role of LRF in (evaluating) alchemical derivatives is stressed, the latter playing a promising role in exploring the chemical compound space. Examples for the transmutation of N_2 and the CC → BN substitution pattern in 2D and 3D carbocyclic systems illustrate the computational efficiency of the use of alchemical derivatives in exploring nearest neighbours in the chemical compound space. As a second perspective, the role of LRF in evaluating and interpreting molecular conductivity is described. Returning to its forerunner, Coulson's atom-atom polarizability, it is shown how in conjugated π systems (and within certain approximations) a remarkable integral-integrand relationship between the atom-atom polarizability and the transmission probability between the atoms/contacts exists, leading to similar trends in both properties. A simple selection rule for transmission probability in alternating hydrocarbons is derived based on the sign of the atom-atom polarizability.

IntechOpen

Keywords: conceptual DFT, linear response function, nearsightedness of electronic matter, alchemical derivatives, molecular conductivity

1. Introduction

A continuous challenge for theoretical and quantum chemists is to see if 'classical' chemical concepts describing bonding, structure and reactivity—the common language of all chemists—can still be retrieved from the nowadays extensive and complex computational results obtained at different levels of complexity with wave function or density functional theory. Conceptual density functional theory (CDFT) [1–6] has played an important role in this endeavour in the past decades. CDFT is a branch of DFT [7, 8] aiming to give precision to often well-known but sometimes vaguely defined chemical concepts (e.g. electronegativity, hardness and softness), affording their numerical evaluation, and to use them either as such or in the context of principles such as Sanderson's electronegativity equalization principle [9] or Pearson's hard and soft acids and bases principle [10]. 'Chemical' DFT or even 'chemical reactivity' DFT would have been a better name for the obvious reason that concepts are essential for all branches of DFT (especially the fundamentals) and that chemical reactivity is one of the main issues addressed in conceptual DFT.

When looking at the basics of CDFT, the energy functional, $E = E[N, v]$, stands out [8]. Why? It is the key ingredient to get (qualitative and quantitative) insight into the eagerness of an atom, or a molecule, to adapt itself to changes in the number of electrons, N, and/or the external potential, $v(\mathbf{r})$, that is, the potential felt by the electrons due to the nuclei. These changes are essential in describing (the onset of) a chemical reaction, hence chemical reactivity. The readiness of a system to adapt itself to these new conditions is quantified through *response functions*, $\delta^m(\partial^n E/\partial N^n)/\delta v(\mathbf{r}_1)\ldots\delta v(\mathbf{r}_m)$, which are the cornerstones of CDFT. Literature on these response functions is abundant, especially on the first-order responses (electronic chemical potential μ [11] $(n = 1, m = 0)$ and the electron density $\rho(\mathbf{r})$ $(n = 0, m = 1)$, the cornerstone of DFT itself) and two of the second-order responses (chemical hardness η $(n = 2, m = 0)$ and its inverse, the chemical softness S [12], and the electronic Fukui function $f(\mathbf{r})(n = 1, m = 1)$ [13]). The most prominent of the third-order response functions [14] is the dual descriptor $f^{(2)}(\mathbf{r})$ [15], the N-derivative of the Fukui function $(n = 1, m = 1)$. Remarkably, response functions diagonal in $v(\mathbf{r})$ $(n = 0, m = 2-3)$ were nearly absent in the CDFT literature until about 10 years ago (see [16] for an overview of this early work). The reasons are obvious; here we concentrate on its simplest member $(n = 0, m = 2)$, the linear response function (LRF).

$$\chi(\mathbf{r}, \mathbf{r}') = \left(\delta^2 E/\delta v(\mathbf{r})\delta v(\mathbf{r}')\right)_N \quad (1)$$

The calculation of this kernel turns out to be far from trivial, as is the representation of this quantity, a function of six Cartesian coordinates, and by extension its link to 'chemical' concepts.

Note that in the context of time-dependent DFT [17–19], the LRF has made its appearance many years ago as it was realized that the poles of its frequency-dependent form are nothing other than the electronic excitation energies. Thanks to Casida's elegant matrix formalism [20], electronic transition frequencies, intensities and assignments are nowadays routinely performed, implemented as they are in standard quantum chemistry packages. However, this evolution was not accompanied by a parallel endeavour on the evaluation, representation and chemical interpretation of the frequency-independent or static LRF.

In the past decade, the ALGC group, in collaboration with colleagues from different countries (Canada (Ayers), US (Yang), Spain (Sola), Poland (Balawender), etc.), set out a program aiming at the systematic evaluation, representation and interpretation of the LRF with the following results obtained until 2013, summarized in a review paper in Chemical Society Reviews [16] (no explicit reference to each of the individual constituting studies will be given).

1. The LRF can now be routinely calculated at several levels of approximation for which the coupled perturbed Kohn-Sham perturbational approach turns out to be the most attractive approach, also permitting different levels of sophistication depending on the treatment of the exchange correlation potential (v_{xc}) term in the perturbation equations. In its simplest form (neglecting the influence of the external potential variation on the Coulomb and exchange-correlation terms in the perturbational equations), the independent particle expression, already presented by Ayers and Parr [8, 21], is retrieved.

2. The representation can be done via contour diagrams (fixing, e.g. r′) as demonstrated for atoms and molecules, or in the case of molecules, after condensation, via a simple atom-atom matrix, reminiscent of reporting the results of a population analysis.

3. An abundance of chemical information was shown to be present in the LRF ranging from the shell structure of atoms, to inductive and mesomeric effects, electron (de)localization and (anti)aromaticity in molecules.

In the present chapter, a synopsis is given of the progress made since then by the ALGC group in collaboration with other groups as witnessed by two of the authors (P.A and R. B.), both on fundamental and applied aspects, that is, on *new insights* into the properties of the LRF and on *new areas* where the LRF is at stake. In Section 2 the mathematical/physical properties of the LRF are revised together with those of its congener, the softness kernel $s(\mathbf{r}, \mathbf{r}')$, the latter playing a fundamental role in scrutinizing Kohn's nearsightedness of electronic matter (NEM) principle. The extension of LRF in the context of spin polarized DFT is also addressed. In Section 3, we highlight the importance of the LRF in the emerging field of alchemical derivatives when exploring the chemical compound space. We illustrate the potential of alchemical derivatives in exploring the CC → BN isoelectronic substitution in 2D and 3D unsaturated carbocyclic molecules: benzene and the C_{60} fullerenes. Finally, in Section 4 we show how the LRF (in fact its forerunner, Coulson's atom-atom polarizability) can be used to predict/interpret the conductivity behaviour of unsaturated hydrocarbons, thus entering the vibrant field of molecular electronics.

2. Theoretical developments

2.1. On the negative and positive semidefiniteness of the LRF and the softness kernel and thermodynamic analogies

The properties of the LRF $\chi(\mathbf{r},\mathbf{r}')$ are intimately related to the concavity/convexity properties of $E[N,v]$ that we addressed in recent years [22, 23]. Where $E[N,v]$ is convex with respect to (w.r.t.) N [24], it is well established that $E[N,v]$ is concave w.r.t. $v(\mathbf{r})$ following the Jensen's inequality.

$$E[N,\lambda v_1 + (1-\lambda)v_2] \geq \lambda E[N,v_1] + (1-\lambda)E[N,v_2] \quad \wedge \ \lambda \in [0,1] \tag{2}$$

(see for example Lieb [25], Eschrig [26], and Helgaker et al. [27]). In **Figure 1** (after Helgaker [28]) we illustrate the physical interpretation of this concavity property. For a given potential v_1, the associated ground state energy is given by the expectation value $\langle\psi_1|H[v_1]|\psi_1\rangle$, with ground state wave function ψ_1, the red point in the figure. Changing v_1 to v_2, with constant ψ, induces changes in energy linear in $\delta v(\mathbf{r})$ as a consequence of the term $\int \rho(\mathbf{r})v(\mathbf{r})d\mathbf{r}$ in the DFT energy expression. Relaxing the wave function yields an energy lowering (blue arrow) until the energy obtained by applying the variational principle with $v = v_2$ is reached (with associated wave function ψ_2). Consequently, the true energy (black line) will always be found below the tangent line ensuring concavity.

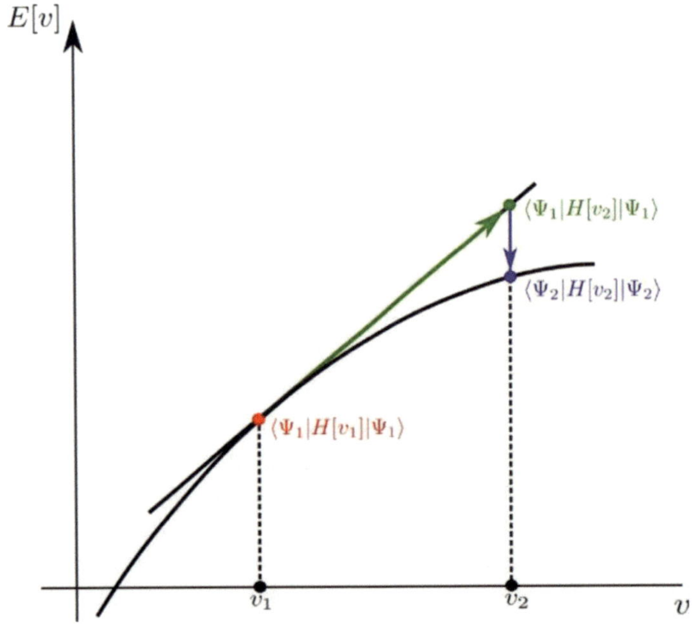

Figure 1. Illustration of the concavity of the $E = E[v]$ functional (after Helgaker [28]) (Reprinted by permission from Springer Nature, Copyright 2016 [22]).

A direct consequence is that the LRF is negative semidefinite.

$$\iint \chi(\mathbf{r},\mathbf{r}')\theta(\mathbf{r})\theta(\mathbf{r}')d\mathbf{r}d\mathbf{r}' \leq 0 \tag{3}$$

where $\theta(\mathbf{r})$ is any continuous function [23]. This inequality shows up when considering the second-order variation of the energy

$$\delta E^{(2)} = 1/2 \iint \left(\delta^2 E/\delta v(\mathbf{r})\delta v(\mathbf{r}')\right)_N \delta v(\mathbf{r})\delta v(\mathbf{r}')d\mathbf{r}d\mathbf{r}' \tag{4}$$

When adopting the $\delta v(\mathbf{r}) = V_0\delta(\mathbf{r}-\mathbf{r}'')$ choice for $\delta v(\mathbf{r})$, where V_0 is a constant one gets

$$\delta E^{(2)} = 1/2 V_0^2 \chi(\mathbf{r}'',\mathbf{r}'') \leq 0 \tag{5}$$

showing that the diagonal elements of the linear response function $\chi(\mathbf{r},\mathbf{r}')$ should be negative or zero. This result links the concavity of the $E[N,v]$ functional to the diagonal elements of the linear response function defined on $R^3 \times R^3$. This negativity was retrieved in all our numerical results (non-integrated and condensed) (see for example [29, 30]). Its physical interpretation is straightforward through the definition of $\chi(\mathbf{r},\mathbf{r}')$ as $(\delta\rho(\mathbf{r})/\delta v(\mathbf{r}'))_N$: if the potential at a given point $\mathbf{r}$ is increased (made less negative), this electron-unfavourable situation will lead to electron depletion at that point, yielding a negative $\chi(\mathbf{r},\mathbf{r})$.

In Section 4, we will point out that Coulson's atom-atom polarizability π_{AB} [31] can be considered as a Hückel-theory forerunner of the LRF and exploit its properties in discussing molecular conductivity. Defined as

$$\pi_{AB} = \partial q_A/\partial\alpha_B = \partial^2 E/\partial\alpha_A\partial\alpha_B = \pi_{BA} \tag{6}$$

the analogy emerges as q_A is the π-electron charge on atom A, whereas (the change in) the Coulomb integral α_B is equivalent to a change in the external potential. The case $A = B$ was explored by Coulson using his complex integral formalism for Hückel's π energy and its derivatives, proving that π_{AA} can be written as

$$\pi_{AA} = (1/\pi)\int_{-\infty}^{+\infty} \left(\Delta_{AA}(iy)/\Delta(iy)\right)^2 dy \tag{7}$$

Here, Δ represents the Hückel secular determinant (or characteristic function) and Δ_{AA} is its counterpart with row A and column A deleted. It turns out that $\Delta_{AA}(iy)/\Delta(iy)$ is imaginary and therefore π_{AA} is negative (see also Section 4).

When introducing the $E = E[N,v]$ functional, it was stressed that in the LRF the second functional derivative is taken at constant N. In the remaining part of this chapter, it turns out that no less important role in CDFT is played by the softness kernel, which is the second functional derivative with respect to $v(\mathbf{r})$ of the $\Omega = \Omega[\mu, v]$ functional (the grand potential) at constant μ. Here, we switch from the canonical ensemble to the grand canonical ensemble [32] connecting two ways of specifying the same physics via functionals of each time two variables differing in one pair that is connected through a Legendre transformation (see for example [33]).

$$\Omega[\mu, v] = E[\tilde{N}[\mu, v], v] - \mu\tilde{N}[\mu, v] \tag{8}$$

where μ and N are conjugate variables which are related by identities $\tilde{N}[\mu, v] = N$ and $\tilde{\mu}[N, v] = \mu$. Its apparent analogy with classical thermodynamics will be addressed later.

In [23], we pointed out that $\Omega[\mu, v]$ just as $E[N, v]$ is concave w.r.t. $v(\mathbf{r})$, implying that its second functional derivative at constant μ, the softness kernel (note the negative sign in the definition) [34].

$$s(\mathbf{r}, \mathbf{r}') = -\left(\delta^2\Omega/\delta v(\mathbf{r})\delta v(\mathbf{r}')\right)_\mu \tag{9}$$

is positive semidefinite. This property fits the well-known Berkowitz-Parr relationship [33] linking $\chi(\mathbf{r}, \mathbf{r}')$ and $s(\mathbf{r}, \mathbf{r}')$ (vide infra).

$$s(\mathbf{r}, \mathbf{r}') = -\chi(\mathbf{r}, \mathbf{r}') + f(\mathbf{r})f(\mathbf{r}')/\eta \tag{10}$$

where $f(\mathbf{r})$ is the Fukui function and η is the hardness. Indeed,

$$\begin{aligned}\iint s(\mathbf{r}, \mathbf{r}')\theta(\mathbf{r})\theta(\mathbf{r}')d\mathbf{r}d\mathbf{r}' &= \iint \left(-\chi(\mathbf{r}, \mathbf{r}') + f(\mathbf{r})f(\mathbf{r}')/\eta\right)\theta(\mathbf{r})\theta(\mathbf{r}')d\mathbf{r}d\mathbf{r}' \\ &= -\iint \chi(\mathbf{r}, \mathbf{r}')\theta(\mathbf{r})\theta(\mathbf{r}')d\mathbf{r}d\mathbf{r}' + 1/\eta\int (f(\mathbf{r})\theta(\mathbf{r}))^2 d\mathbf{r} \geq 0\end{aligned} \tag{11}$$

since the hardness is nonnegative and $\chi(\mathbf{r}, \mathbf{r}')$ was shown to be negative semidefinite.

The properties of $\chi(\mathbf{r}, \mathbf{r}')$ and $s(\mathbf{r}, \mathbf{r}')$ incited us to reconsider the analogy between the DFT functionals E and Ω [32] on the one hand and the macroscopic thermodynamic state functions $U = U[S, V]$, $F = F[T, V]$, $H = H[S, P]$ and $G = G[T, P]$ on the other hand (internal energy, Helmholtz free energy, enthalpy and Gibbs free energy written as functions of volume (V), temperature (T), pressure (P) and entropy (S)). Parr and Nalewajski extended the notion of intensive and extensive variables (T, P) and (S, V), respectively, in thermodynamics to the variables in DFT functionals by classifying external variables as properties additive with respect to any partitioning of the electron density $\rho(\mathbf{r}) = \rho_A(\mathbf{r}) + \rho_B(\mathbf{r})$ [32]. In this way, $\rho(\mathbf{r})$ and N are clearly extensive, and μ and $v(\mathbf{r})$ are intensive. The analogy between $G[T, P]$ and $\Omega[\mu, v]$ can now be stressed: both the state function G and the DFT functional Ω contain two intensive variables. This situation leads to a remarkable property when formulating a DFT analogue of the stability analysis in macroscopic thermodynamics [33, 35, 36]. Concavity for $G[T, P]$ in all directions then implies that

$$d^2G = G_{TT}(\Delta T)^2 + 2G_{TP}\Delta T\Delta P + G_{PP}(\Delta P)^2 \leq 0, \tag{12}$$

where G_{TT}, G_{TP} and G_{PP} are the second derivatives of Gibbs free energy, $(G_{XY} = \partial^2 G/\partial X \partial Y)$. This stability condition implies the negative semidefiniteness of the Hessian matrix. It yields

$$G_{PP} = (\partial V/\partial P)_T = -\kappa_T V \leq 0 \tag{13}$$

$$G_{TT} = -(\partial S/\partial T)_P = -C_P/T \leq 0 \tag{14}$$

that is, the isothermal compressibility κ_T and the heat capacity at constant pressure C_P should be positive, and the condition that

$$\left(G_{TT}G_{PP} - G_{PT}^2\right) = \left(\kappa_T V C_P/T - \alpha^2 V^2\right) \geq 0 \Leftrightarrow \kappa_T C_P/T \geq \alpha^2 V \tag{15}$$

where $\alpha = (\partial V/\partial T)_P/V$ is the coefficient of thermal expansion. Another classical example of such stability analysis is the entropy written as $S[U, V]$ (at constant number of particles) where both variables are now extensive [33].

Let us now consider the analogy with $\Omega[\mu, v]$. It is well known that

$$\left(\partial^2\Omega/\partial\mu^2\right)_v = -S[\mu, v], \tag{16}$$

where S is the global softness [3]. As the r.h.s. of (16) is negative, concavity for $\Omega[\mu, v]$ in μ shows up. As discussed above, Ω is also concave in $v(\mathbf{r})$, leading to the positive semidefiniteness of $s(\mathbf{r}, \mathbf{r}')$ (these expressions being the counterparts of (13) and (14)). The condition for concavity in all directions leads after some algebra (see [23]) to the condition

$$\begin{aligned}&\left(\left(\partial^2\Omega/\partial\mu^2\right)_v \Delta\mu + \int \left(\partial(\delta\Omega/\delta v(\mathbf{r}))_\mu/\partial\mu\right)_v \Delta v(\mathbf{r})d\mathbf{r}\right)^2 + \iint \left(\left(\partial^2\Omega/\partial\mu^2\right)_v \left(\delta^2\Omega/\delta v(\mathbf{r})\delta v(\mathbf{r}')\right)_\mu\right.\\ &\quad \left. - \left(\partial(\delta\Omega/\delta v(\mathbf{r}))_\mu/\partial\mu\right)_v \left(\partial(\delta\Omega/\delta v(\mathbf{r}'))_\mu/\partial\mu\right)_v\right)\Delta v(\mathbf{r})d\mathbf{r}\Delta v(\mathbf{r}')d\mathbf{r}' \geq 0\end{aligned} \tag{17}$$

and finally to

$$\iint (s(\mathbf{r}, \mathbf{r}')S - s(\mathbf{r})s(\mathbf{r}'))\Delta v(\mathbf{r})d\mathbf{r}\Delta v(\mathbf{r}')d\mathbf{r}' \geq 0, \tag{18}$$

the analogue of (15) where the local softness

$$s(\mathbf{r}) = \left(\partial(\delta\Omega/\delta v(\mathbf{r}))_\mu/\partial\mu\right)_v = (\partial\rho(\mathbf{r})/\partial\mu)_v \tag{19}$$

has been introduced [37]. Taking again for $\Delta v(\mathbf{r})$ and $\Delta v(\mathbf{r}')$ Dirac delta functions $\delta(\mathbf{r} - \mathbf{r}'')$ and $\delta(\mathbf{r}' - \mathbf{r}'')$, one obtains the condition

$$s(\mathbf{r}, \mathbf{r})S \geq (s(\mathbf{r}))^2 \geq 0. \tag{20}$$

This inequality shows that the diagonal elements $s(\mathbf{r}, \mathbf{r})$ should be positive, as could be inferred from the concavity of $\Omega[\mu, v]$ but, more importantly, they impose a restriction on the relative values of the three softness descriptors $s(\mathbf{r},\mathbf{r}')$,$S$ and $s(\mathbf{r})$ in analogy to the thermodynamic relationship between κ_T, C_P and α at given T and P ($V = V[T,P]$). This result is compatible with the aforementioned Berkowitz-Parr relationship. Indeed, starting from their expression for $\mathbf{r}' = \mathbf{r}$ (see (10), the relations $s(\mathbf{r}) = f(\mathbf{r})/\eta$)

$$s(\mathbf{r},\mathbf{r}) = -\chi(\mathbf{r},\mathbf{r}) + s(\mathbf{r})s(\mathbf{r})/S \tag{21}$$

and knowing that $\chi(\mathbf{r},\mathbf{r}) \leq 0$ (vide supra) one obtains

$$s(\mathbf{r},\mathbf{r}) - s(\mathbf{r})s(\mathbf{r})/S \geq 0, \tag{22}$$

retrieving our conclusions above.

2.2. Kohn's nearsightedness of electronic matter revisited

We now report on our recent explorations [38] on Kohn's NEM principle. Kohn introduced the NEM concept in 1996 [39] and elaborated on it in 2005 with Prodan et al. [40]. In his own words, it can be viewed as 'underlying such important ideas as Pauling's chemical bond, transferability, and Yang's computational principle of divide and conquer' [40]. Certainly in view of the two former issues, this principle, formulated by a physicist, touches the very heart of chemistry and so, in our opinion, it was tempting to look at it with a chemist's eye. Why however is this issue addressed in this chapter; in other words, what is the link between the LRF and nearsightedness?

The quintessence of the NEM principle is as follows (**Figure 2**): consider a (many) electron system characterized by an electron density function $\rho(\mathbf{r})$ with a given electronic chemical potential μ. Now, concentrate on a point $\mathbf{r}_0$ and perturb the system in its external potential $v(\mathbf{r})$ at point $\mathbf{r}'$, outside a sphere with radius R around $\mathbf{r}_0$ at constant electronic potential μ. Then, the NEM principle states that the absolute value of the density change at $\mathbf{r}_0$, $|\Delta\rho(\mathbf{r}_0)|$, will be lower than a finite maximum value $\overline{\Delta\rho}$, which depends on $\mathbf{r}_0$ and R, whatever the magnitude of the perturbation. As stated by Kohn, anthropomorphically the particle density $\rho(\mathbf{r})$cannot 'see' any perturbation $v(\mathbf{r})$ beyond the distance $R\left[\mathbf{r}_0, \overline{\Delta\rho}\right]$ within an accuracy $\overline{\Delta\rho}$, that is, the density shows nearsightedness. Kohn offered evidence that in the case of 1D 'gapped' systems (i.e. with hardness η larger than zero) the decay of $\overline{\Delta\rho}$ as a function of R (i.e. upon increasing $|\mathbf{r} - \mathbf{r}_0|$) is exponential and that for gapless systems it follows a power law. This suggests that in the molecular world, where η is observed to be always positive, the electron density should only be sensitive to nearby changes in the external potential. In [38], we provided the first numerical confirmation of this nearsightedness principle for real, 3D, molecules.

Again, why address this issue in this LRF chapter? Going back to Kohn's formulation quintessentially a change in density at a given point $\Delta\rho(\mathbf{r}_0)$ in response to a change in external potential $\Delta v(\mathbf{r}')$ at different points is analysed. These are the typical ingredients of the LRF (change in v produces a change in ρ), and in this case the process is considered at constant electronic chemical potential μ, in order words $(\delta\rho(\mathbf{r})/\delta v(\mathbf{r}'))_\mu$ is the key quantity. This is nothing else than the softness kernel $s(\mathbf{r},\mathbf{r}')$ with a minus sign in front. Indeed,

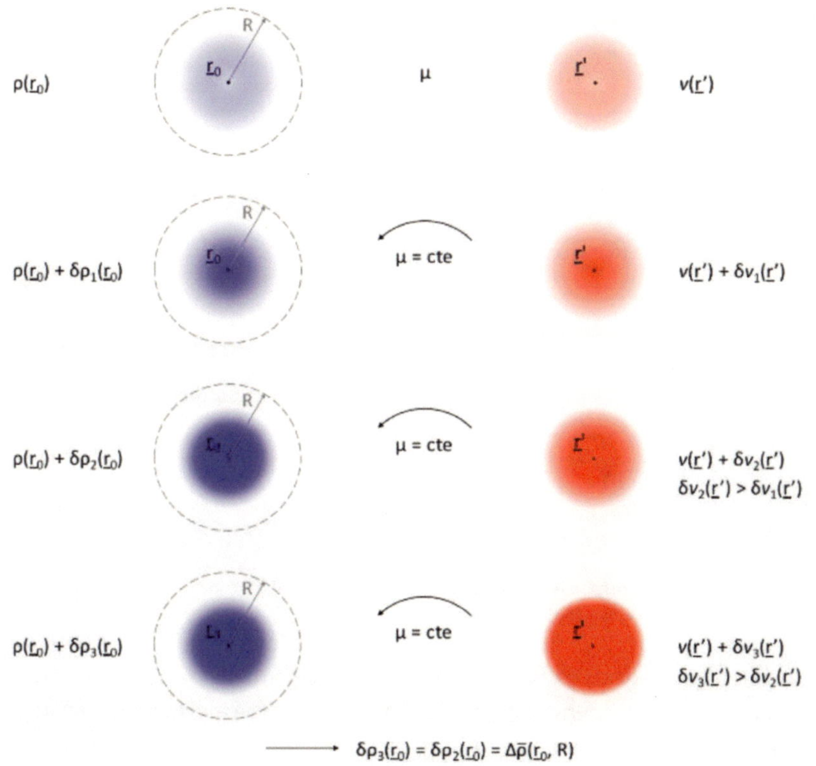

Figure 2. Pictorial representation of the nearsightedness of electronic matter principle: when $\delta v(\mathbf{r}') \geq \delta v_2(\mathbf{r}')$, with $\mathbf{r}'$ outside a sphere with radius R around $r_0, \delta\rho_0(r_0)$ no longer increases—$\delta\rho_0(\mathbf{r}_0) = \overline{\Delta\rho}(\mathbf{r}_0, R)$ no matter how large $\delta v(\mathbf{r}')$. The intensity of the red and blue regions represents the magnitude of $v(\mathbf{r}')$ and $\rho(\mathbf{r})$, respectively.

$$(\delta\rho(\mathbf{r})/\delta v(\mathbf{r}'))_\mu = -s(\mathbf{r},\mathbf{r}') = \left(\delta^2\Omega/\delta v(\mathbf{r})\delta v(\mathbf{r}')\right)_\mu \tag{23}$$

The Berkowitz-Parr relationship [34] can then be written as

$$(\delta\rho(\mathbf{r})/\delta v(\mathbf{r}'))_N = (\delta\rho(\mathbf{r})/\delta v(\mathbf{r}'))_\mu - \left(\partial\rho(\mathbf{r})/\partial\mu\right)_v \left(\delta\mu/\delta v(\mathbf{r}')\right)_N \tag{24}$$

an equation transforming conditions of constant N into constant μ, for taking the functional derivative of $\rho(\mathbf{r})$ w.r.t. $\delta v(\mathbf{r})$ in analogy with this type of equation for partial derivatives in macroscopic thermodynamics [33]. As $\left(\partial\rho(\mathbf{r})/\partial\mu\right)_v$ equals the local softness $s(\mathbf{r})$ and $\left(\delta\mu/\delta v(\mathbf{r}')\right)_N$ is an alternative way, as compared to $(\partial\rho(\mathbf{r})/\partial N)_v$ to write the Fukui function f(r), one retrieves from (28)

$$s(\mathbf{r},\mathbf{r}') = -\chi(\mathbf{r},\mathbf{r}') + f(\mathbf{r})f(\mathbf{r}')/\eta \tag{25}$$

which will be the key equation in this section. As analytical methods are available to evaluate $\chi(\mathbf{r},\mathbf{r}'), f(\mathbf{r})$ and η on equal footing [41], $s(\mathbf{r},\mathbf{r}')$ can be evaluated, and its difference with $\chi(\mathbf{r},\mathbf{r}')$ analysed, in order to scrutinize nearsightedness.

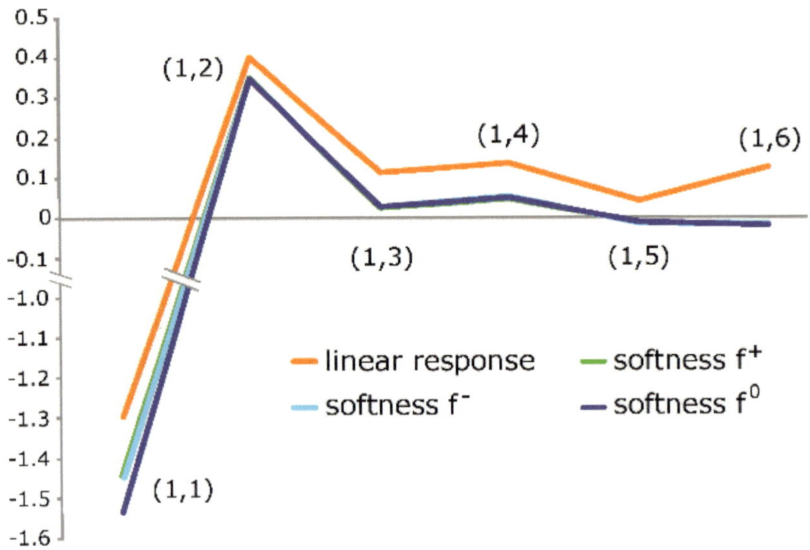

Figure 3. Atom condensed linear response function and softness kernel of 1,3,5-hexatriene. The curves of the softness kernels using f+ and f− are overlapping [38].

In **Figure 3**, we depict the atom condensed linear response function and the softness kernel, the matrices χ_{AB} and s_{AB}, of 1,3,5-hexatriene. Condensation was performed by integrating the kernels over the domains of atoms A and B, represented by V_A and V_A, that is,

$$s_{AB} = \int\limits_{V_A}\int\limits_{V_B} s(\mathbf{r},\mathbf{r}')d\mathbf{r}d\mathbf{r}' \tag{26}$$

From our previous work on polyenes [42], the χ_{AB} matrix elements are known to show an alternating behaviour with maxima on mesomeric active atoms (C_2,C_4 and C_6) and minima for mesomeric passive atoms (C_3 and C_5), with the change in $v(\mathbf{r})$ taking place at C_1.The picture illustrates that the softness kernel is more nearsighted than the LRF (all its values are lower) and that $s_{1,5}$ and $s_{1,6}$ are very close to zero; the effect of the perturbation has died off completely, confirming the nearsightedness of the softness kernel. This effect can be traced back to a cancellation of the LRF, which is non-nearsighted, by the second term in Eq. (25), and accounts for the density changes induced by charge transfer from the electron reservoir to keep the chemical potential constant. Note that this condition, at first sight somewhat strange, is an often more realistic perspective when considering for example the reactivity of molecules in solution where the chemical potential is fixed by the solvent allowing (partial) charge transfer to or from the molecule while keeping its chemical potential constant [43, 44].

As a second example, we show in **Figure 4** the change in density of the 1,3,5-heptatrienyl cation when the C atom of one of the terminal CH_2 atoms was alchemically replaced by an N atom. The corresponding density difference was evaluated through the alchemical derivatives approach (Section 3) where the carbon atom of the CH_2 group was annihilated and replaced by a nitrogen atom at constant geometry and constant number of electrons. It is clear that the

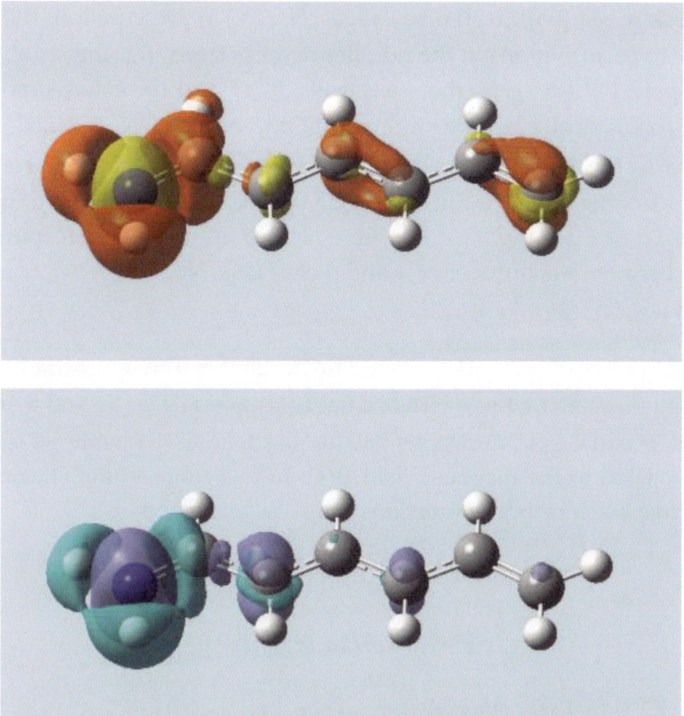

Figure 4. Alchemical change in density using the linear response (top) and softness kernel (bottom) for the heptatrienyl cation to 1,3,5-hexatriene-1-amine [38].

effect of functionalization dies off much more quickly under constant μ conditions. In fact, the third carbon atom is hardly affected in this—it should be stressed—unsaturated system, offering the possibility for mesomerism. In the LRF, the decay is slow and effects are still relatively important five bonds away from the perturbation.

Further case studies on '3D' systems (e.g. alchemically changing methylcubane to fluorocubane and on functionalized neopentane) yield similar results. All together, the results on the near-sightedness of the softness kernel found for all systems discussed in [38] are the first and a firm numerical confirmation of Kohn's NEM principle in the molecular world. To put it in chemical terms, these findings provide computational evidence for the transferability of functional groups: molecular systems can be divided into locally interacting subgroups retaining a similar functionality and reactivity that can only be influenced by changes in the direct environment of the functional group. Thus, the physicist's NEM principle and the chemist's transferability principle [45]—at the heart of, for example, the whole of organic chemistry [46]—are reconciled.

2.3. Extension of $\chi(\mathbf{r}, \mathbf{r}')$ in the context of spin polarized CDFT

As a natural extension of CDFT to the case where spin polarization is included [47–49], spin polarized conceptual DFT was introduced by Galvan et al. [50, 51] (see also Ghanty and Ghosh

[52] and for a review see [53]). In the so-called (N_α, N_β) representation, the response of the electronic energy to perturbations in the number of α electrons, N_α, the number of β electrons, N_β, the external potential acting on the α electrons, $v_\alpha(\mathbf{r})$, and the external potential acting on the β electrons, $v_\beta(\mathbf{r})$, is studied. In this representation, the $E = E[N, v]$ functional from Section 2.1 is generalized to the $E = E[N_\alpha, N_\beta, v_\alpha, v_\beta]$ functional. In the second, equivalent, (N, N_S) representation, in fact the one introduced by Galvan, the $E = E[N, v]$ functional is generalized to $E = E[N, N_S, v, v_S]$ or $E = E[N, N_S, v, \mathbf{B}]$ where N_S denotes the electron spin number defined as the difference between the number of α and β electrons, $N_S = N_\alpha - N_\beta$, whereas v_s is given by $v_S(\mathbf{r}) = (v_\alpha(\mathbf{r}) - v_\beta(\mathbf{r}))/2$.

$v(\mathbf{r})$ is equal to$(v_\alpha(\mathbf{r}) + v_\beta(\mathbf{r}))/2$ and $\mathbf{B}$ represents an external magnetic field. If the magnetic field is static and uniform along the z axis, one has $v_\alpha(\mathbf{r}) = v(\mathbf{r}) + \mu_B B_Z$ and $v_\beta(\mathbf{r}) = v(\mathbf{r}) - \mu_B B_Z$, $v(\mathbf{r})$ is the usual external spin-free potential (as used in CDFT without spin polarization), whereas $v_S(\mathbf{r})$ is related to the magnetic field $\mathbf{B}(\mathbf{r})$. In this context, and sticking to the (N, N_S) representation, three linear response functions can now be defined:

$$\chi_{NN}(\mathbf{r}, \mathbf{r}') = \left(\delta^2 E/\delta v(\mathbf{r})\delta v(\mathbf{r}')\right)_{N, N_S, v_S} \tag{27}$$

$$\chi_{SS}(\mathbf{r}, \mathbf{r}') = \left(\delta^2 E/\delta v_S(\mathbf{r})\delta v_S(\mathbf{r}')\right)_{N, N_S, v} \tag{28}$$

$$\begin{aligned} \chi_{NS}(\mathbf{r}, \mathbf{r}') &= (\delta^2 E/\delta v(\mathbf{r})\delta v_S(\mathbf{r}'))_{N, N_S} = (\delta \rho_N(\mathbf{r})/\delta v_S(\mathbf{r}'))_{N, N_S, v} \\ &= (\delta \rho_S(\mathbf{r}')/\delta v(\mathbf{r}))_{N, N_S, v} = \left(\delta^2 E/\delta v_S(\mathbf{r}')\delta v(\mathbf{r})\right)_{N, N_S} = \chi_{SN}(\mathbf{r}', \mathbf{r}) \end{aligned} \tag{29}$$

where $\rho_N(\mathbf{r}) = \rho_\alpha(\mathbf{r}) + \rho_\beta(\mathbf{r})$ and $\rho_S(\mathbf{r}) = \rho_\alpha(\mathbf{r}) - \rho_\beta(\mathbf{r})$ are the total and spin densities, respectively. $\chi_{NN}(\mathbf{r}, \mathbf{r}')$ is the analogue of the spin-independent CDFT expression for the LRF $\chi(\mathbf{r}, \mathbf{r}')$ (Eq. (1)).

Space limitations prevent us to go in detail on the results reported in [47, 48]. We only depict in **Figure 5** the SPCDFT analogue of the contour plots for $\chi(\mathbf{r}, \mathbf{r}')$ for closed shell atoms as discussed in [47] and the Chem Soc Rev paper [16]. In **Figure 5**, we plot the LRF, in the (N_α, N_β) representation, for the ground state of Li ($(1s)^2$ $(2s)^1$). The structure of $\chi_{\alpha\alpha}(\mathbf{r}, \mathbf{r}')$ and $\chi_{\beta\beta}(\mathbf{r}, \mathbf{r}')$ is similar to the $\chi(\mathbf{r}, \mathbf{r}')$ plots for closed shell atoms: a negative diagonal part (cfr [47] and Section 2.1) and alternating positive and negative parts for r or r′ = constant in order to obey the trivial equation (cf. [47])

$$\int \chi(\mathbf{r}, \mathbf{r}')d\mathbf{r} = \int (\delta \rho_N(\mathbf{r})/\delta v(\mathbf{r}'))_N d\mathbf{r} = (\delta N/\delta v(\mathbf{r}'))_N = 0 \tag{30}$$

$\chi_{\alpha\alpha}$ extends further away from the nucleus than $\chi_{\beta\beta}$ in line with the extra α electron with the higher principal quantum number (n = 2) extending farther away from the nucleus than the n = 1 β electron. $\chi_{\beta\beta}$ looks similar to the $\chi_{\beta\beta}$ plot for He (insert) but contracted more to the origin (due to higher nuclear charge). The $\chi_{\alpha\beta}$ and $\chi_{\alpha\beta}$ plots show some evident symmetry, $\chi_{\alpha\beta}(\mathbf{r}, \mathbf{r}') = \chi_{\beta\alpha}(\mathbf{r}', \mathbf{r})$, and show positive regions along the diagonal. A positive perturbation in

http://dx.doi.org/10.5772/intechopen.80280

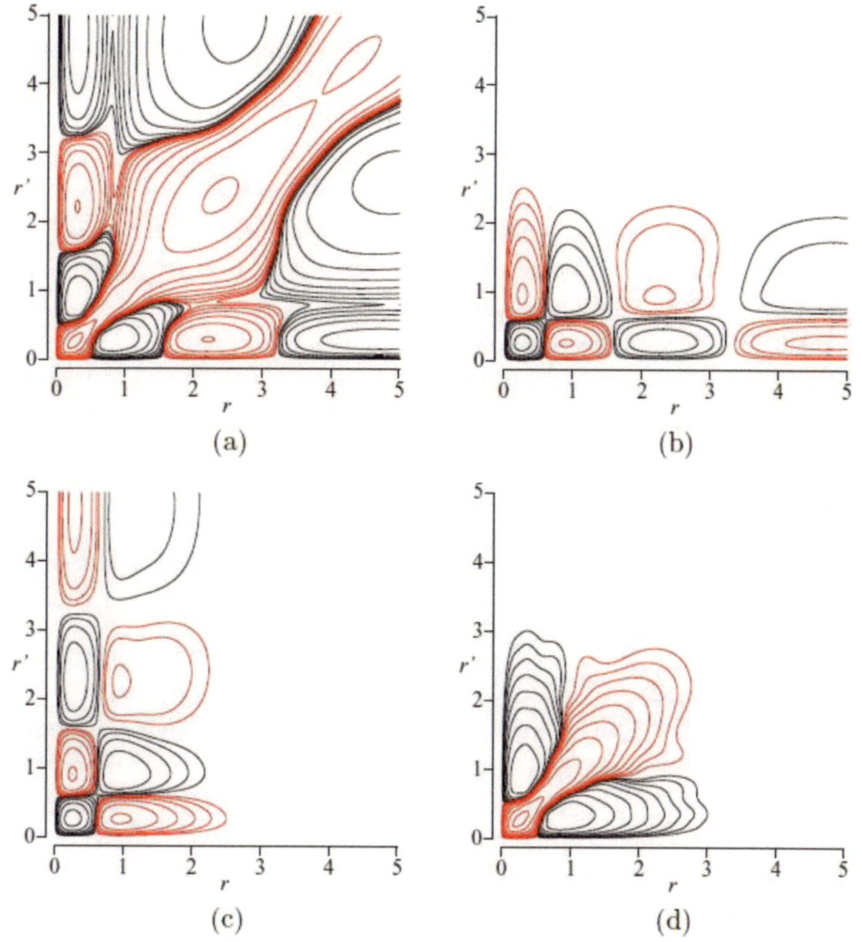

Figure 5. Contour plots of the radial distribution function of the spin polarized linear response function of Lithium in the [Nα, Nβ] representation. r is represented on the horizontal axis, r′ on the vertical axis [(a) Lithium $\chi_{\alpha\alpha}$, (b) Lithium $\chi_{\alpha\beta}$, (c) Lithium $\chi_{\alpha\alpha}$, (d) Lithium $\chi_{\beta\beta}$. In the insert, the $\chi_{\beta\beta}$ plot for He (see text) (Reprinted from [48] with the permission of AIP Publishing)].

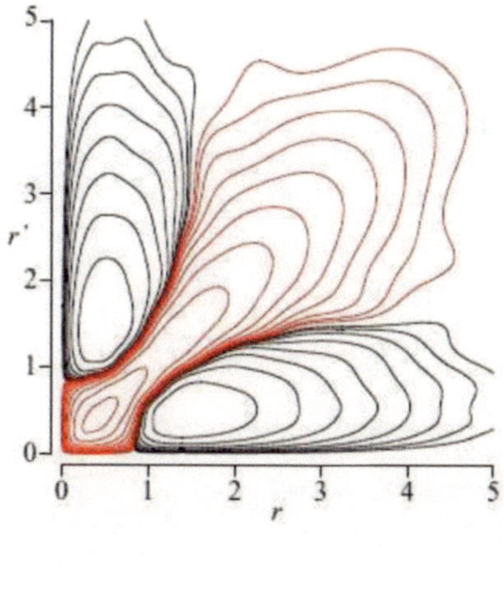

the α external potential $\delta v_\alpha(\mathbf{r})$ will cause a depletion of electrons in the vicinity of the perturbation, the β electrons are not affected directly. However, the depletion in α electrons will influence the Coulomb potential and due to the lower electron-electron repulsion, an accumulation in β electrons in the region considered will occur resulting in a positive diagonal $\chi_{\beta\alpha}(\mathbf{r},\mathbf{r})$ value. The concentration of the $\chi_{\alpha\beta}$ and $\chi_{\alpha\beta}$ isocontours along the r and r′ axes can be interpreted when referring to the $\chi_{\alpha\alpha}$ and $\chi_{\beta\beta}$ plots: perturbing $v_\beta(\mathbf{r})$ at a distance r′ larger than 3 a.u. clearly has no effect on the β density, the Coulomb potential, the overall density and consequently on the α density. This results in $(\delta\rho_\alpha(\mathbf{r})/\delta v_\beta(\mathbf{r}')) = \chi_{\alpha\beta}(\mathbf{r},\mathbf{r}')$ having zero values for r′ larger than 3 a.u. On the other hand, perturbing $v_\beta(\mathbf{r})$ close to the nucleus induces a change in the β density, with repercussion on the Coulomb potential and so on the α density farther away from the nucleus (on the r axis) even in regions where the β density is no longer affected. All these features account for the 'partial plane filling' of the $\chi_{\alpha\beta}$ and $\chi_{\alpha\beta}$ plots with a 'demarcation' line at 3 a.u.

To close this section, we mention that once $\chi(\mathbf{r},\mathbf{r}')$ (or its counterparts in SPCDFT) is known, a local version of the polarizability tensor components, say α_{xy}, namely $\alpha_{xy}(\mathbf{r})$ can be obtained by straightforward integration:

$$\alpha_{xy}(\mathbf{r}) = -\int x(\mathbf{r})\chi(\mathbf{r},\mathbf{r}')y(\mathbf{r}')d\mathbf{r}' \tag{31}$$

An example is given in **Figure 6** [48] where for the atoms Li through Ne the trend of the spherically averaged $\alpha(\mathbf{r})$ $\left(\frac{1}{3}\left(\alpha_{xx}(r) + \alpha_{yy}(r) + \alpha_{zz}(r)\right)\right)$ is given. From 2 a.u. on, the trends in $\alpha(\mathbf{r})$ for Li up to Ne parallel the global polarizability, known to decrease along a period of the periodic table. At lower distances (preceding the valence region), inversions with even negative values occur. The present results are evidently important when, for example, disentangling reaction mechanisms where local polarizabilities, that is, in certain regions of the reagents, are

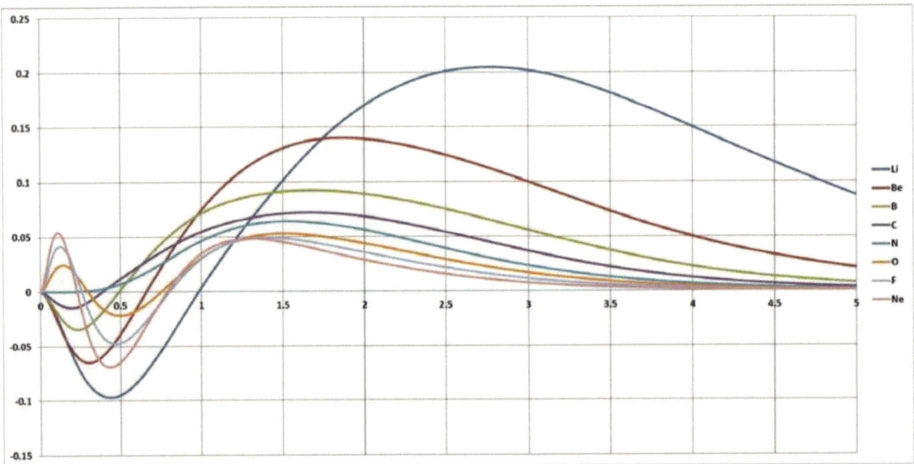

Figure 6. Plot of the local polarizability α(r) of the atoms Li through Ne via CPKS (see text) (Reprinted by permission of the publisher (Taylor and Francis Ltd.) [48]).

at stake and not the overall polarizability. The parallel between the relationship between local and global softness (Section 2.1) is obvious.

3. The role of the LRF in alchemical derivatives and exploring chemical compound space

3.1. Context

The LRF has recently been exploited when investigating Chemical Compound Space [54–56]. Chemists are continuously exploring chemical compound space (CCS) [57, 58], the space populated by all imaginable chemicals with natural nuclear charges and realistic interatomic distances for which chemical interactions exist. Navigating through this space is costly, obviously for synthetic-experimental chemists and also for theoretical and computational chemists who might and should be guides for indicating relevant domains in CCS to their experimental colleagues. Doing even a simple single-point SCF calculation at every imaginable point leads to prohibitively large computing times (not to speak about bookkeeping aspects and manipulation of the computed data). A very promising ansatz was initiated by Von Lilienfeld et al. [59–63] in his alchemical coupling approach where two, isoelectronic molecules in CCS are coupled 'alchemically' through the interpolation of their external potentials (see also the work by Yang and co-workers on designing molecules by optimizing potentials [64]). At the heart of this ansatz are the alchemical derivatives, partial derivatives of the energy w.r.t. one or more nuclear charges at constant number of electrons and geometry. The simplest members of this new family of response functions are:

the alchemical potential

$$\mu_A^{\mathrm{al}}[N,\mathbf{Z},\mathbf{R}] \equiv (\partial W/\partial Z_A)_{\mathcal{N},\mathbf{R}} = (\partial E/\partial Z_A)_{\mathcal{N},\mathbf{R}} + (\partial V_{\mathrm{nn}}/\partial Z_A)_{\mathcal{N},\mathbf{R}} = \mu_A^{\mathrm{al,el}} + \mu_A^{\mathrm{al,nuc}} \tag{32}$$

and the alchemical hardness

$$\eta_{AB}^{\mathrm{al}}[N,\mathbf{Z},\mathbf{R}] \equiv (\partial^2 W/\partial Z_A \partial Z_B)_{\mathcal{N},\mathbf{R}} = (\partial^2 E/\partial Z_A \partial Z_B)_{\mathcal{N},\mathbf{R}} + (\partial^2 V_{\mathrm{nn}}/\partial Z_A \partial Z_B)_{\mathcal{N},\mathbf{R}} = \eta_{AB}^{\mathrm{al,el}} + \eta_{AB}^{\mathrm{al,nuc}} = \eta_{BA}^{\mathrm{al}}, \tag{33}$$

where $W[N,\mathbf{Z},\mathbf{R}] \equiv E[N,v[\mathbf{Z},\mathbf{R}]] + V_{\mathrm{nn}}[\mathbf{Z},\mathbf{R}]$ is the total energy of a system,$\mathbf{R} = (\mathbf{R}_A, \mathbf{R}_B, \ldots)$ denotes constant geometry and $\mathbf{Z} = (Z_A, Z_B, \ldots,)$denotes the nuclear charges vector. The analogy with the electronic chemical potential and hardness (see Section 1) is striking. Some years ago, one of the present authors (R.B.) presented a strategy to calculate these derivatives for any atom or atom-atom combination analytically and to use them, starting from a single SCF calculation on the parent or reference molecule, to explore the CCS of first neighbours, implying changes of nuclear charges of +1 or −1 [65]. In this way, simple arithmetic can be used, starting from a Taylor expansion

$$\Delta W[d\mathbf{Z}] = W[N,\mathbf{Z}+d\mathbf{Z},\mathbf{R}] - W[N,\mathbf{Z},\mathbf{R}] = \sum_A \mu_A^{\mathrm{al}} dZ_A + 1/2 \sum_A \sum_B \eta_{AB}^{\mathrm{al}} dZ_A dZ_B \tag{34}$$

instead of a new SCF calculation for each transmutant.

The position of the alchemical derivatives in CDFT was already mentioned: they are response functions, now related to a particular charge in external potential, namely the charge in one or more nuclear charges. As the second derivatives are 'diagonal' in these particular external potential changes, a direct link with the LRF can be expected. Using the chain rule, one easily writes

$$\begin{aligned}\eta_{AB}^{\text{al, el}} = \left(\partial^2 E[N, v[\mathbf{Z}, \mathbf{R}]]/\partial Z_A \partial Z_B\right)_{N,\mathbf{R}} &= \iint (\delta\rho(\mathbf{r})/\delta v(\mathbf{r}'))_N (\partial v(\mathbf{r})/\partial Z_A)_{N,\mathbf{R}} (\partial v(\mathbf{r}')/\partial Z_B)_{N,\mathbf{R}} d\mathbf{r}d\mathbf{r}' \\ &= \iint \chi(\mathbf{r}, \mathbf{r}')(1/|\mathbf{r} - \mathbf{R}_A|)(1/|\mathbf{r}' - \mathbf{R}_B|) d\mathbf{r}d\mathbf{r}'\end{aligned} \tag{35}$$

indicating that the alchemical hardness is obtained by integration of the LRF after multiplication by $1/|\mathbf{r} - \mathbf{R}_A|$ and $1/|\mathbf{r}' - \mathbf{R}_B|$. The basic relationship of the alchemical derivatives and the LRF shows how, again, the LRF makes its appearance in sometimes unexpected areas (see also Section 4). In the case of the first derivative, the $(\delta E/\delta v(\mathbf{r}))_N$ factor in the integrand simplifies to $\rho(\mathbf{r})$(see Section 1), yielding.

$$\mu_A^{\text{al, el}} = \int \rho(\mathbf{r})/|\mathbf{r} - \mathbf{R}_A| d\mathbf{r} \tag{36}$$

the electronic potential at the nucleus, well known as the electronic part of the molecular electrostatic potential [66].

3.2. Applications

As a very simple example, we consider the transmutation of the nitrogen molecule. Five chemically relevant mutants can be generated (see **Figure 7**) as nearest neighbours in CCS ($\Delta Z = \pm 1$) and at constant number of electrons: CO, NO^+, O_2^{2+}, CN^- and C_2^{2-}. The differences between 'vertical' (i.e. exact, via two SCF calculations) and alchemical transmutation energies were evaluated at the B3LYP/cc-pCVTZ level (note the inclusion of additional tight functions 'C in

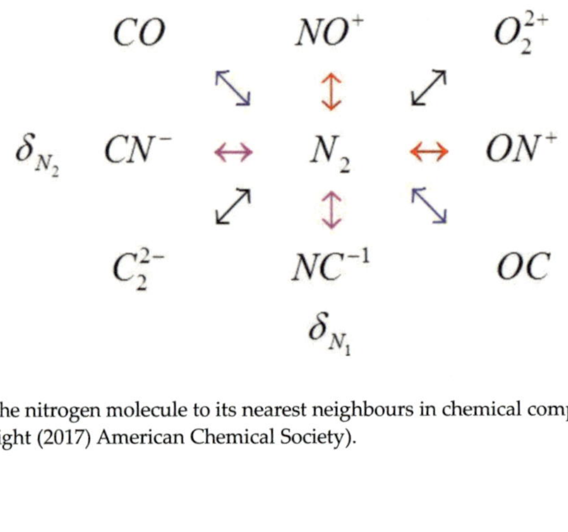

Figure 7. Transmutation of the nitrogen molecule to its nearest neighbours in chemical compound space (Reprinted with permission from [54]. Copyright (2017) American Chemical Society).

order to recover the (changes in) core-core and core-valence correlation). The mean absolute error was only 0.034 a.u., the $N_2 \rightarrow$ COtransmutation being particularly successful with a difference in energy of only 0.004 a.u., that is, 2 kcal mol $^{-1}$. The transmutation into a neutral system results in the cancellation of the odd terms in the Taylor expansion. Of course, we do not claim 'chemical accuracy', yet the ordering of the energy of all transmutants came out correctly, indicating that the alchemical procedure (even when stopped at second order) is a simple, straightforward road to explore CCS for neighbouring structures.

Similar conclusions could be drawn for transmutation of benzene, for example, by the substitution of CC units by their isoelectronic BN units. The replacement of a CC unit in an aromatic molecule by an isoelectronic unit BN has been shown to impart important, interesting electronic, photophysical and chemical properties, often distinct from the parent hydrocarbon [67]. An in-depth study of all azoborines (**Figure 8**) $C_6H_6 \rightarrow C_{6\text{-}2n}H_6(BN)_n$ ($n = 1, 2, 3$) turned out to reproduce correctly the stability of all possible isomers for a given n value (3,11,3 for $n = 1, 2, 3$), which is of importance for applications in graphene chemistry where the $(CC)_n \rightarrow (BN)_n$ substitution is a topic that has received great interest in recent years [68].

As a computational 'tour de force' , and passing from '2D' benzene to '3D' fullerenes, we recently explored the alchemical approach to study the complete $(CC)_n \rightarrow (BN)_n$ substitution pattern of C_{60}, all the way down to $(BN)_{30}$. $C_{60} \rightarrow C_{60\text{-}2n}(BN)_n$, $n = 1, 2, \ldots, 30$, predicting and interpreting via 'alchemical rules' the most stable isomers for each value of n. This study is based on a *single* SCF calculation on C_{60} and its alchemical derivatives up to second order, enabling each possible transmutation energy to be evaluated by simple arithmetic (the diagonal elements of the alchemical hardness matrix are equal)

$$\Delta W_n[\mathbf{N} = (N1, \ldots, Nn), \mathbf{B} = (B1, \ldots, Bn)] = n\,\eta_{11}^{\mathrm{al}} + \sum_{i=1} \sum_{j>i} \left(\eta_{\mathrm{N}i\mathrm{N}j}^{\mathrm{al}} + \eta_{\mathrm{B}i\mathrm{B}j}^{\mathrm{al}}\right) - \sum_{i=1} \sum_{j=1} \eta_{\mathrm{N}i\mathrm{B}j'}^{\mathrm{al}} \quad (37)$$

where ΔW_n is the transmutation energy from C_{60} to $C_{60\text{-}2n}(BN)_n$ and $\mathbf{N}/\mathbf{B}$ is a vector of the carbon atoms replaced by the nitrogen/boron atoms.

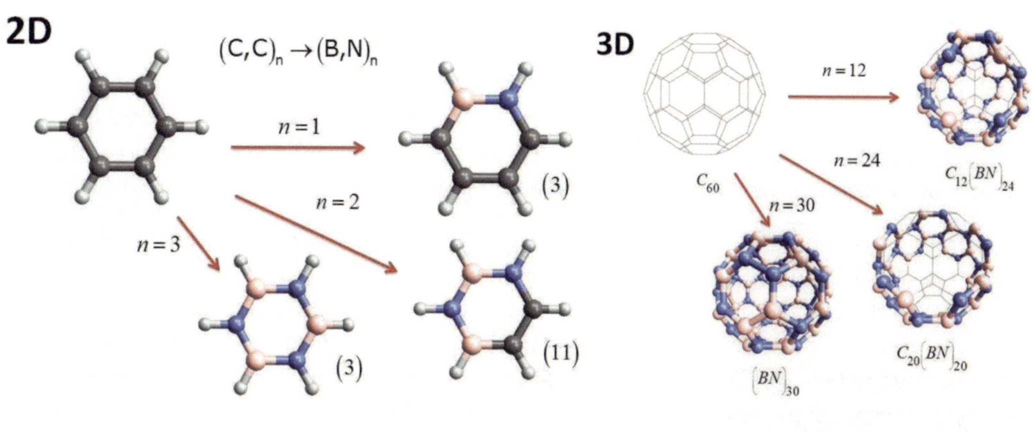

Figure 8. CC-BN Substitutions in 2D and 3D unsaturated carbocyclic systems (number of isomers for the 2D case in parentheses).

In (37), it is seen that the linear term drops as μ_i is unique by symmetry for all atoms in C_{60} and because $\Sigma \Delta Z_i = 0$ for any transmutation. The alchemical hardness matrix $\boldsymbol{\eta}$ thus completely determines the substitution energy and pattern. To summarize the results (for an in-depth discussion see [55]), the study reveals that the correct sequence of stabilization energies for each n is retrieved by adopting an approach in which for each n value the problem is looked upon without prejudice of the n − 1 result (called the 'simultaneous' approach). The other, simpler method (we called it the 'successive' approach) was shown to fail already after n = 13, be it that at some n values identical values were obtained with both approaches, for example, for the Belt structure (n = 20). Needless to say that even the 'successive' approach might already be prohibitively demanding for standard ab initio or DFT calculations, let it be for the simultaneous method. The sequence of substitutions could be interpreted in terms of a number of 'alchemical' rules of which the two most important are when (referring to earlier work by Kar et al. [69]): (1°) hexagon-hexagon junctions are preferably substituted, in a way that minimizes the homonuclear (BB,NN) bonds and (2°) the higher stability is created by maximizing the number of filled hexagons. For the subsequent, more intricate, rules we refer to [55].

Shortage of space prevents us to comment on our recent results on the evaluation of isolated atom alchemical derivatives up to third order with different techniques, from numerical differentiation (not discussed hitherto in this chapter), via the coupled perturbed Kohn-Sham approach as discussed before, to the March and Parr combined $1/Z$ and $N^{-1/3}E[N, Z]$ expansion ansatz [70], permitting a walk through the periodic table on the road to scrutinize periodicity effects [56].

4. The role of the LRF (or its forerunner, the atom-atom polarizability) in evaluating and interpreting molecular electronics

4.1. Context

In this section the role of the LRF in molecular electronics is highlighted [72] has been a vibrant area of research in recent years. An ever-increasing number of papers (both experimental and theoretical) studied the transport properties of typically organic molecules containing π-conjugated systems and considering possible applications for incorporation in molecular electronic devices (MED) [73]. Most of these theoretical studies have been performed at a high level of theory, but this type of calculation does not always lead to simple insights into why some molecules will conduct and which will insulate, and how the positions of the contacts influence this behaviour.

We therefore adopted a simple ansatz based on Ernzerhof's source and sink potential (for details see [74, 75]) in Fowler's tight-binding Hückel approach [76]. One thereby considers only the π electrons of the molecule and cuts the resonance effects between the contact and the molecule after the molecule's nearest neighbour in the contact. In the so-called weak interaction limit (see details in [71]), the transmission probability at the Fermi level $T(0)$, which is directly proportional to the conductance, can then be written as

$$T(0) = 4\beta^2 \Delta_{AB}^2 / \Delta^2 \quad (38)$$

Here, as in Section 2.1, Δ is the Hückel determinant for the isolated molecule, from which in Δ_{AB} its A-th row and B-th column are deleted and β is a measure of the interaction between the contact atom and its molecular neighbour. Again, what is the role of the LRF in this road to calculating/understanding molecular conductivity?

4.2. The atom-atom polarizability as a conductivity indicator

Going back to expression (7) for the diagonal elements of Coulson's atom-atom polarizability, a general element π_{AB} of this forerunner of the LRF can be written as

$$\pi_{AB} = (1/\pi)\int_{-\infty}^{+\infty} (\Delta_{AB}(iy)/\Delta(iy))^2 dy \tag{39}$$

(the contour integral in the complex plane in Coulson's formalism [31] is hereby reduced to an integral along the imaginary axis). The integrand of π_{AB} thus turns out to be related to the transition probability at the Fermi level T_r $T_{AB}(0)$ when the contacts are placed at atoms A and B of the molecule. An intimate relation between π_{AB} and $T_{AB}(0)$ thus exists; the precise connection was evaluated by explicit calculation of π_{AB} and $T_{AB}(0)$, some results being visualized in **Figure 9**. There, we depict both the π_{AB} and $T_{AB}(0)$ values for some linear acenes (benzene, naphthalene, anthracene and tetracene), taking always one atom as the reference atom. For the atom-atom polarizability, the reference atom is denoted by a green circle with its area proportional to the self-polarizability of that atom which is as pointed out in Section 2.1 as always negative. Black and red circles on the other atoms denote the atom-atom polarizability values corresponding to a perturbation on the reference atom (varying α_A) on the (charge of the) considered atom $B(q_B)$. Black circles correspond to $\pi_{AB} > 0$, red circles to $\pi_{AB} < 0$. For $T_{AB}(0)$, a similar approach is followed. The empty green circle denotes the position of the first contact 'A' (zero 'ipso' transmission [71]), the magnitude of the black circles on the other atoms 'B' denotes the magnitude of the transmission when the second contact is placed at position B; note that no red circles arise as $T_{AB}(0)$, the transmission probability, always lies between 0 and 1.

Figure 9 shows that the pattern in the two plots is completely analogous and leads to the conjecture that a positive atom-atom polarizability seems to be a necessary condition in these Kekulean benzenoids in order to have transmission for a certain configuration of the contacts on the molecule. If the areas of the circles are considered, no exact proportionality between π_{AB} and $T_{AB}(0)$ can be inferred, but they are correlated.

This issue was further investigated in the next linear acene and pentacene (**Figure 10**), by varying the position of the first contact. For a fixed first contact, the highest transmission occurs when the second contact is at the atom with the highest π_{AB} value involving the first atom. From that atom on, a monotonously decreasing probability is noticed in either directions. The sharpest decline in T is in the direction of the neighbour that exhibits the lowest atom-atom polarizability. On the basis of these results, it can indeed be conjectured that a positive atom-atom polarizability is a necessary condition for transmission and that the tendencies between the two series of values are similar.

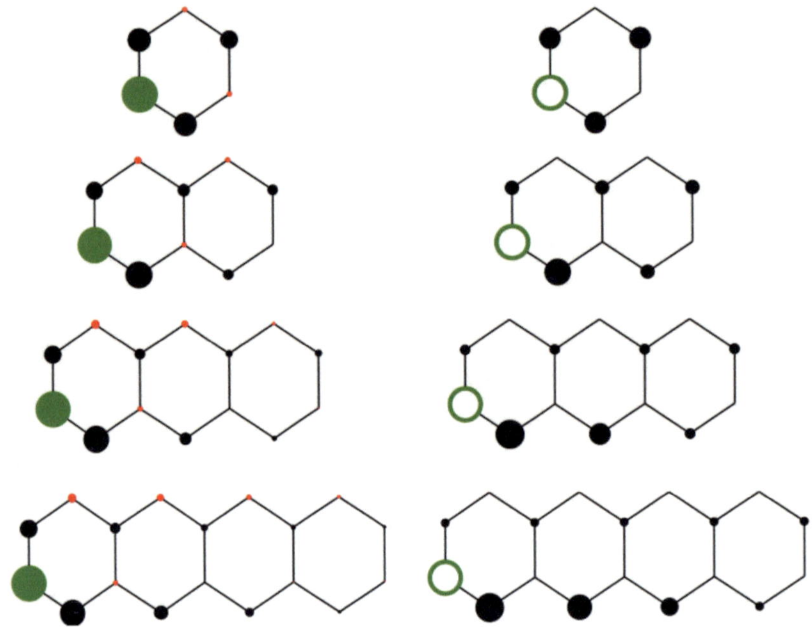

Figure 9. The atom-atom polarizability (left) and transmission probability at the Fermi level (right) for a single reference atom for benzene, naphthalene, anthracene and tetracene (Reprinted from [71] with permission of AIP Publishing).

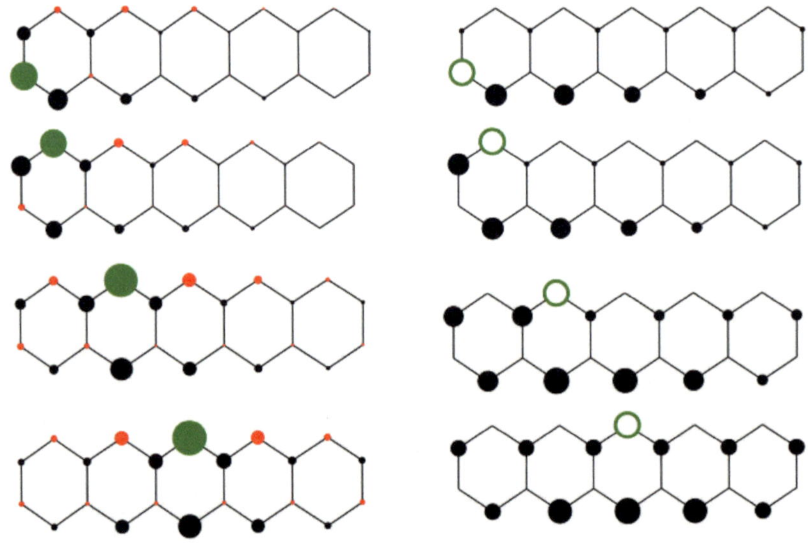

Figure 10. The atom-atom polarizability (left) and transmission probability at the Fermi level (right) for pentacene for variable reference atom (Reprinted from [71] with permission of AIP Publishing).

Further analysis of the behaviour of the $(\Delta_{AB}/\Delta)^2$ function in the complex plane was done in the case of alternant non-singular hydrocarbons (for details see [71]). Coulson's and Longuet Higgins' pairing theorem shows that if A and B are drawn from the same partite set (all 'starred' or 'unstarred' atoms, respectively) $\Delta^2_{AB}(z)/\Delta(z)^2$ is odd and if A and B are taken from a different set this function is even. It is then easily seen that the odd function has no real part along the imaginary axis, leading to a negative $(\Delta_{AB}/\Delta)^2$ value along the y axis and zero at the origin, yielding a negative atom-atom polarizability and zero transmission at the Fermi level ($T(0)$). On the other hand, when A and B are drawn from opposite sets, Δ_{AB}/Δ is an even function yielding real values along the y axis. A positive π_{AB} value results with either insulation or transmission, depending on whether $\Delta_{AB}(0)$ is zero or not. Note that the analysis for $A = B$ leads to the same result as for $A \neq B$ but belonging to the same set yielding a negative π_{AA} value. The following overall conclusion, formulated as a selection rule, can be drawn: negative atom-atom polarizabilities for non-singular alternant hydrocarbons correspond to devices with insulation at the Fermi level. Conduction at the Fermi level requires, but is not guaranteed by, a positive atom-atom polarizability.

The aforementioned properties were used as guiding principle in our later studies towards a chemical interpretation of molecular electronic conductivity [77] leading to a simple, back-of-the-envelope determination of quantum interference [78], thus bridging the gap between chemical reactivity theory and molecular electronics.

5. Conclusions

The LRF and its congener, the softness kernel, are now in a stage where many of their mathematical and physical properties are well understood. The possibility to evaluate, represent and interpret them puts them on equal footing for their use in conceptual DFT with their already more traditional second-order companions, the chemical hardness and the Fukui function. In view of the 'chemistry' contained in the LRF kernel as shown some years ago, it is not unexpected, but it still remains to be unravelled whether they are major players in very fundamental issues pertaining to the electronic structure of matter as in Kohn's nearsightedness of electronic matter principle, as well as in more applied fields where they are shown to be of great use to explore chemical compound space (through the alchemical derivatives) and to predict/interpret molecular conductivity.

Acknowledgements

P.G. and F.D.P. acknowledge the Vrije Universiteit Brussel (VUB) for a Strategic Research Program. F.D.P. also acknowledges the Franqui Foundation for a position as Francqui Research Professor. S.F. acknowledges the Research Foundation Flanders (FWO) and the European Union's Horizon 2020 Marie Sklodowska-Curie grant (N° 706415) for financially supporting his postdoctoral research at the ALGC group. T.S. acknowledges the FWO for a position as research assistant (11ZG615N). P.W.A. thanks the Natural Sciences and Engineering Research Council, the Canada Research Chairs and Compute Canada for financial support. R.B. thanks the Interdisciplinary Centre for Mathematical and Computational Modelling for a computing grant. P.G. thanks Kristina Nikolova for her meticulous help in styling the manuscript and Tom Bettens for his help in styling **Figure 2**.

Conflicts of interest

.The authors report no conflicts of interest regarding this publication.

Author details

Paul Geerlings[1]*, Stijn Fias[1,2], Thijs Stuyver[1], Paul Ayers[2], Robert Balawender[3] and Frank De Proft[1]

*Address all correspondence to: pgeerlin@vub.be

1 Algemene Chemie, Vrije Universiteit Brussel, Brussels, Belgium

2 Department of Chemistry and Chemical Biology, McMaster University, Hamilton, ON, Canada

3 Institute of Physical Chemistry, Polish Academy of Sciences, Warsaw, Poland

References

[1] Parr RG, Yang W. Density-functional theory of the electronic structure of molecules. Annual Review of Physical Chemistry. 1995;**46**:701-728. DOI: 10.1146/annurev.pc.46.100195.003413

[2] Chermette H. Chemical reactivity indexes in density functional theory. Journal of Computational Chemistry. 1999;**20**:129-154. DOI: 10.1002/(SICI)1096-987X(19990115)20:1<129::AID-JCC13>3.3.CO;2-1

[3] Geerlings P, De Proft F, Langenaeker W. Conceptual density functional theory. Chemical Reviews. 2003;**103**:1793-1874. DOI: 10.1021/cr990029p

[4] Ayers PW, Anderson JSM, Bartolotti LJ. Perturbative perspectives on the chemical reaction prediction problem. International Journal of Quantum Chemistry. 2005;**101**:520-534. DOI: 10.1002/qua.20307

[5] Gázquez JL. Perspectives on the density functional theory of chemical reactivity. Journal of the Mexican Chemical Society. 2008;**52**:3-10

[6] Liu SB. Conceptual density functional theory and some recent developments. Acta Physico-Chimica Sinica. 2009;**25**:590-600. DOI: 10.3866/PKU.WHXB20090332

[7] Hohenberg P, Kohn W. Inhomogeneous electron gas. Physical Review. 1964;**136**:B864-B871. DOI: 10.1103/PhysRev.136.B864

[8] Parr RG, Weitao Y. Density-Functional Theory of Atoms and Molecules. International Series of Monographs on Chemistry. London/New York: Oxford University Press; 1989

[9] Sanderson RT. An interpretation of bond lengths and a classification of bonds. Science. 1951;**114**:670-672. DOI: 10.1126/science.114.2973.670

[10] Pearson RG. Hard and soft acids and bases. Journal of the American Chemical Society. 1963;**85**:3533-3539. DOI: 10.1021/ja00905a001

[11] Parr RG, Donnelly RA, Levy M, Palke WE. Electronegativity: The density functional viewpoint. The Journal of Chemical Physics. 1978;**68**:3801-3807. DOI: 10.1063/1.436185

[12] Parr RG, Pearson RG. Absolute hardness: Companion parameter to absolute electronegativity. Journal of the American Chemical Society. 1983;**105**:7512-7516. DOI: 10.1021/ja00364a005

[13] Parr RG, Yang W. Density functional approach to the frontier-electron theory of chemical reactivity. Journal of the American Chemical Society. 1984;**106**:4049-4050. DOI: 10.1021/ja00326a036

[14] Geerlings P, De Proft F. Conceptual DFT: The chemical relevance of higher response functions. Physical Chemistry Chemical Physics. 2008;**10**:3028. DOI: 10.1039/b717671f

[15] Morell C, Grand A, Toro-Labbé A. New dual descriptor for chemical reactivity. The Journal of Physical Chemistry. A. 2005;**109**:205-212. DOI: 10.1021/jp046577a

[16] Geerlings P, Fias S, Boisdenghien Z, De Proft F. Conceptual DFT: Chemistry from the linear response function. Chemical Society Reviews. 2014;**43**:4989. DOI: 10.1039/c3cs60456j

[17] Runge E, Gross EKU. Density-functional theory for time-dependent systems. Physical Review Letters. 1984;**52**:997-1000. DOI: 10.1103/PhysRevLett.52.997

[18] Marques MAL, Ullrich CA, Nogueira F, Rubio A, Burke K, Gross EKU. Time-Dependent Density Functional Theory. Lecture Notes in Physics. Vol. 706. Berlin, Heidelberg: Springer; 2006. DOI: 10.1007/b11767107

[19] Marques MAL, Maitra NT, Nogueira FMS, Gross EKU, Rubio A. Fundamentals of Time-Dependent Density Functional Theory. Lecture Notes in Physics. Vol. 837. Berlin, Heidelberg: Springer; 2012. DOI: 10.1007/978-3-642-23518-4

[20] Casida ME. Time-dependent density functional response theory for molecules. In: Recent Advances in Density Functional Methods. Singapore: World Scientific; 1995. pp. 155-192. DOI: 10.1142/9789812830586_0005

[21] Ayers PW. Strategies for computing chemical reactivity indices. Theoretical Chemistry Accounts. 2001;**106**:271-279. DOI: 10.1007/PL00012385

[22] Geerlings P, Boisdenghien Z, De Proft F, Fias S. The E = E[N, v] functional and the linear response function: A conceptual DFT viewpoint. Theoretical Chemistry Accounts. 2016; **135**:213. DOI: 10.1007/s00214-016-1967-9

[23] Geerlings P, De Proft F, Fias S. Analogies between density functional theory response kernels and derivatives of thermodynamic state functions. Acta Physico-Chimica Sinica. 2018;**34**:699-707

[24] Perdew JP, Parr RG, Levy M, Balduz JL. Density-functional theory for fractional particle number: Derivative discontinuities of the energy. Physical Review Letters. 1982;**49**:1691-1694. DOI: 10.1103/PhysRevLett.49.1691

[25] Lieb EH. Density functionals for Coulomb systems. International Journal of Quantum Chemistry. 1983;**24**:243-277. DOI: 10.1002/qua.560240302

[26] Eschrig H. The Fundamentals of Density Functional Theory. Stuttgart-Leipzig, Germany: Teubner; 1996

[27] Kvaal S, Ekström U, Teale AM, Helgaker T. Differentiable but exact formulation of density-functional theory. Journal of Chemical Physics. 2014;**140**:18A518. DOI: 10.1063/1.4867005

[28] Helgaker T. In: 13th Sostrup Summer School on Quantum Chemistry and Molecular Properties; 2014

[29] Sablon N, De Proft F, Ayers PW, Geerlings P. Computing second-order functional derivatives with respect to the external potential. Journal of Chemical Theory and Computation. 2010;**6**:3671-3680. DOI: 10.1021/ct1004577

[30] Boisdenghien Z, Van Alsenoy C, De Proft F, Geerlings P. Evaluating and interpreting the chemical relevance of the linear response kernel for atoms. Journal of Chemical Theory and Computation. 2013;**9**:1007-1015. DOI: 10.1021/ct300861r

[31] Coulson CA, Longuet-Higgins HC. The electronic structure of conjugated systems. I. General theory. Proceedings of the Royal Society of London. Series A: Mathematical and Physical Sciences. 1947;**191**:39-60. DOI: 10.1098/rspa.1947.0102

[32] Nalewajski RF, Parr RG. Legendre transforms and Maxwell relations in density functional theory. The Journal of Chemical Physics. 1982;**77**:399-407. DOI: 10.1063/1.443620

[33] Callen HB. Thermodynamics and an Introduction to Thermostatistics. New York, USA: Wiley; 1985

[34] Berkowitz M, Parr RG. Molecular hardness and softness, local hardness and softness, hardness and softness kernels, and relations among these quantities. The Journal of Chemical Physics. 1988;**88**:2554-2557. DOI: 10.1063/1.454034

[35] Prigogine I, Defay R. Chemical Thermodynamics. London, UK: Longmans; 1954

[36] Berry RS, Rice SA, Ross J. Physical Chemistry. New York, USA: John Wiley & Sons; 1980

[37] Yang W, Parr RG. Hardness, softness, and the Fukui function in the electronic theory of metals and catalysis. Proceedings of the National Academy of Sciences. 1985;**82**:6723-6726. DOI: 10.1073/pnas.82.20.6723

[38] Fias S, Heidar-Zadeh F, Geerlings P, Ayers PW. Chemical transferability of functional groups follows from the nearsightedness of electronic matter. Proceedings of the National Academy of Sciences. 2017;**114**:11633-11638. DOI: 10.1073/pnas.1615053114

[39] Kohn W. Density functional and density matrix method scaling linearly with the number of atoms. Physical Review Letters. 1996;**76**:3168-3171. DOI: 10.1103/PhysRevLett.76.3168

[40] Prodan E, Kohn W. Nearsightedness of electronic matter. Proceedings of the National Academy of Sciences. 2005;**102**:11635-11638. DOI: 10.1073/pnas.0505436102

[41] Yang W, Cohen AJ, De Proft F, Geerlings P. Analytical evaluation of Fukui functions and real-space linear response function. The Journal of Chemical Physics. 2012;**136**:144110. DOI: 10.1063/1.3701562

[42] Sablon N, De Proft F, Geerlings P. The linear response kernel: Inductive and resonance effects quantified. Journal of Physical Chemistry Letters. 2010;**1**:1228-1234. DOI: 10.1021/jz1002132

[43] Cárdenas C, Rabi N, Ayers PW, Morell C, Jaramillo P, Fuentealba P. Chemical reactivity descriptors for ambiphilic reagents: Dual descriptor, local hypersoftness, and electrostatic potential. The Journal of Physical Chemistry. A. 2009;**113**:8660-8667. DOI: 10.1021/jp902792n

[44] Cárdenas C, Echegaray E, Chakraborty D, Anderson JSM, Ayers PW. Relationships between the third-order reactivity indicators in chemical density-functional theory. The Journal of Chemical Physics. 2009;**130**:244105. DOI: 10.1063/1.3151599

[45] Bader RFW. Nearsightedness of electronic matter as seen by a physicist and a chemist. The Journal of Physical Chemistry. A. 2008;**112**:13717-13728. DOI: 10.1021/jp806282j

[46] Patai S. PATAI's chemistry of functional groups. Wiley Online. DOI: 10.1002/SERIES1078

[47] Boisdenghien Z, Fias S, Van Alsenoy C, De Proft F, Geerlings P. Evaluating and interpreting the chemical relevance of the linear response kernel for atoms II: Open shell. Physical Chemistry Chemical Physics. 2014;**16**:14624. DOI: 10.1039/c4cp01331j

[48] Fias S, Boisdenghien Z, De Proft F, Geerlings P. The spin polarized linear response from density functional theory: Theory and application to atoms. The Journal of Chemical Physics. 2014;**141**:184107. DOI: 10.1063/1.4900513

[49] Boisdenghien Z, Fias S, Da Pieve F, De Proft F, Geerlings P. The polarizability of atoms and molecules: A comparison between a conceptual density functional theory approach and time-dependent density functional theory. Molecular Physics. 2015;**113**:1890-1898. DOI: 10.1080/00268976.2015.1021110

[50] Galván M, Gázquez JL, Vela A. Fukui function: Spin-density and chemical reactivity. The Journal of Chemical Physics. 1986;**85**:2337-2338. DOI: 10.1063/1.451083

[51] Galvan M, Vela A, Gazquez JL. Chemical reactivity in spin-polarized density functional theory. The Journal of Physical Chemistry. 1988;**92**:6470-6474. DOI: 10.1021/j100333a056

[52] Ghanty TK, Ghosh SK. Spin-polarized generalization of the concepts of electronegativity and hardness and the description of chemical binding. Journal of the American Chemical Society. 1994;**116**:3943-3948. DOI: 10.1021/ja00088a033

[53] De Proft F, Chamorro E, Pérez P, Duque M, De Vleeschouwer F, Geerlings P. Spin-polarized reactivity indices from density functional theory: Theory and applications. Chemical Modelling. 2009;**6**:63-111. DOI: 10.1039/b812888j. Cambridge: Royal Society of Chemistry

[54] Balawender R, Welearegay MA, Lesiuk M, De Proft F, Geerlings P. Exploring chemical space with the alchemical derivatives. Journal of Chemical Theory and Computation. 2013;**9**:5327-5340. DOI: 10.1021/ct400706g

[55] Balawender R, Lesiuk M, De Proft F, Geerlings P. Exploring chemical space with alchemical derivatives: BN-simultaneous substitution patterns in C 60. Journal of Chemical Theory and Computation. 2018;**14**:1154-1168. DOI: 10.1021/acs.jctc.7b01114

[56] Balawender R, Holas A, De Proft F, Van Alsenoy C, Geerlings P. Alchemical Derivatives of Atoms: A Walk Through the Periodic Table. Many-body Approaches at Different Scales. Cham, Switzerland: Springer International Publishing; 2018. pp. 227-251. DOI: 10.1007/978-3-319-72374-7_20

[57] Kirkpatrick P, Ellis C. Chemical space. Nature. 2004;**432**:823-823. DOI: 10.1038/432823a

[58] Dobson CM. Chemical space and biology. Nature. 2004;**432**:824-828. DOI: 10.1038/nature03192

[59] von Lilienfeld OA, Tuckerman ME. Molecular grand-canonical ensemble density functional theory and exploration of chemical space. The Journal of Chemical Physics. 2006;**125**:154104. DOI: 10.1063/1.2338537

[60] von Lilienfeld OA, Tuckerman ME. Alchemical variations of intermolecular energies according to molecular grand-canonical ensemble density functional theory. Journal of Chemical Theory and Computation. 2007;**3**:1083-1090. DOI: 10.1021/ct700002c

[61] von Lilienfeld OA. Accurate ab initio energy gradients in chemical compound space. The Journal of Chemical Physics. 2009;**131**:164102. DOI: 10.1063/1.3249969

[62] von Lilienfeld OA. First principles view on chemical compound space: Gaining rigorous atomistic control of molecular properties. International Journal of Quantum Chemistry. 2013;**113**:1676-1689. DOI: 10.1002/qua.24375

[63] Chang KYS, Fias S, Ramakrishnan R, von Lilienfeld OA. Fast and accurate predictions of covalent bonds in chemical space. The Journal of Chemical Physics. 2016;**144**:174110. DOI: 10.1063/1.4947217

[64] Wang ML, Hu XQ, Beratan DN, Yang WT. Designing molecules by optimizing potentials. Journal of the American Chemical Society. 2006;**128**:3228-3232. DOI: 10.1021/ja0572046

[65] Lesiuk M, Balawender R, Zachara J. Higher order alchemical derivatives from coupled perturbed self-consistent field theory. The Journal of Chemical Physics. 2012;**136**:034104. DOI: 10.1063/1.3674163

[66] Bonaccorsi R, Scrocco E, Tomasi J. Molecular SCF calculations for the ground state of some three-membered ring molecules: $(CH_2)_3$, $(CH_2)_2NH$, $(CH_2)_2NH_2^+$, $(CH_2)_2O$, $(CH_2)_2S$, $(CH)_2CH_2$, and N_2CH_2. The Journal of Chemical Physics. 1970;**52**:5270-5284. DOI: 10.1063/1.1672775

[67] Campbell PG, Marwitz AJV, Liu S-Y. Recent advances in azaborine chemistry. Angewandte Chemie, International Edition. 2012;**51**:6074-6092. DOI: 10.1002/anie.201200063

[68] Kan M, Li Y, Sun Q. Recent advances in hybrid graphene-BN planar structures. Wiley Interdisciplinary Reviews: Computational Molecular Science. 2016;**6**:65-82. DOI: 10.1002/wcms.1237

[69] Kar T, Pattanayak J, Scheiner S. Rules for BN-substitution in BCN−fullerenes. Separation of BN and C domains. The Journal of Physical Chemistry. A. 2003;**107**:8630-8637. DOI: 10.1021/jp035744o

[70] March NH, Parr RG. Chemical potential, Teller's theorem, and the scaling of atomic and molecular energies. Proceedings of the National Academy of Sciences. 1980;**77**:6285-6288. DOI: 10.1073/pnas.77.11.6285

[71] Stuyver T, Fias S, De Proft F, Fowler PW, Geerlings P. Conduction of molecular electronic devices: Qualitative insights through atom-atom polarizabilities. The Journal of Chemical Physics. 2015;**142**:094103. DOI: 10.1063/1.4913415

[72] Aviram A, Ratner MA. Molecular rectifiers. Chemical Physics Letters. 1974;**29**:277-283. DOI: 10.1016/0009-2614(74)85031-1

[73] Sun L, Diaz-Fernandez YA, Gschneidtner TA, Westerlund F, Lara-Avila S, Moth-Poulsen K. Single-molecule electronics: From chemical design to functional devices. Chemical Society Reviews. 2014;**43**:7378-7411. DOI: 10.1039/C4CS00143E

[74] Goyer F, Ernzerhof M, Zhuang M. Source and sink potentials for the description of open systems with a stationary current passing through. The Journal of Chemical Physics. 2007;**126**:144104. DOI: 10.1063/1.2715932

[75] Ernzerhof M. A simple model of molecular electronic devices and its analytical solution. The Journal of Chemical Physics. 2007;**127**:204709. DOI: 10.1063/1.2804867

[76] Pickup BT, Fowler PW. An analytical model for steady-state currents in conjugated systems. Chemical Physics Letters. 2008;**459**:198-202. DOI: 10.1016/j.cplett.2008.05.062

[77] Stuyver T, Fias S, De Proft F, Geerlings P. Back of the envelope selection rule for molecular transmission: A curly arrow approach. Journal of Physical Chemistry C. 2015;**119**:26390-26400. DOI: 10.1021/acs.jpcc.51310395

[78] Solomon GC, Andrews DQ, Van Duyne RP, Ratner MA. When things are not as they seem: Quantum interference turns molecular electron transfer "rules" upside down. Journal of the American Chemical Society. 2008;**130**:7788-7789. DOI: 10.1021/ja801379b

Section 2

Applications

Chapter 2

Modeling with DFT and Chemical Descriptors Approach for the Development of Catalytic Alloys for PEMFCs

Alejandro E. Pérez and Rafael Ribadeneira

Additional information is available at the end of the chapter

http://dx.doi.org/10.5772/intechopen.80922

Abstract

Material properties and process modeling with density functional theory (DFT) is an accurate method to facilitate the study and the design of materials computationally for the development of different electrochemical technologies such as fuel cells, solar cells, and batteries, among others, mainly to achieve alternative ways for energy conversion and storage. Considering the relevance of DFT in the development of these alternative technologies for energy generation and storage, in this chapter, the application of DFT to study catalytic alloys and their reactivity processes to develop polymer membrane fuel cells (PEMFCs) is presented. In this sense, firstly, a brief review of the application of DFT to develop catalysts for PEMFCs and the relation with the concept of chemical descriptors is presented. Secondly, the main chemical descriptors for this task are presented and discussed. Finally, a summary of the main findings of the modeling with DFT and chemical descriptors approach of catalytic alloys for PEMFCs is presented and analyzed.

Keywords: DFT, chemical descriptors, catalytic alloys, PEMFCs

1. Introduction

The rational design of materials is an important and challenging task, which implies the in-depth understanding of determining factors for the desired properties of the materials. These properties emerge from the material chemistry (atoms involved and its electronic structure) and material structure (organization at nano and microscale) [1, 2], for which the understanding at atomistic level is key and could enable the tailoring of materials that improve the efficiency of almost any technology. Particularly, catalyst development is crucial for efficient chemical production and clean energy generation and storage technologies such as photoelectrochemical cells, fuel cells, and batteries, which are required for an environmentally responsible economic growth [3].

In the last few decades, density functional theory (DFT) has been successfully used to study heterogeneous catalysis process and the properties of the catalysts, achieving an understanding of reaction mechanisms and the determining factors in the catalytic activity of the materials. For example, it has been found that the electronic structure of the catalytic surface determines its properties [4–6]. Also, some general relations between these factors and the catalytic activity have been developed and used for the search and design of catalytic materials for different applications [7–14].

In the next sections, the application of DFT to study catalytic materials and their reactivity to develop polymer membrane fuel cells (PEMFCs) is presented. The first section is devoted to review briefly and in a general way, the different works, where DFT is applied for studying catalysts for the electrodes of PEMFCs and some determining factors on catalysis found in some of these works are drawn. Secondly, the main factors used for the development of catalysts through the concept of chemical descriptors are presented and discussed. Finally, a summary of the main findings of the modeling with DFT and chemical descriptors approach of catalytic alloys for PEMFCs is presented and analyzed.

2. Using DFT to study heterogeneous catalysis in PEMFCs

Fuel cells are devices that convert chemical energy of supplied fuel and oxidant into electrical energy. Their operation involves two electrodes, an anode and a cathode, separated by an electrolyte. The anode fuel, such as hydrogen, methanol, and ethanol, which have been proposed for PEMFCs, is oxidized and donates electrons. Electrons travel through the extern circuit, where they perform work and reach the cathode, where oxidant, generally oxygen, is reduced combining with electrons. The electrolyte, which in the case of PEMFCs is a conducting polymer, conducts electricity through the movement of ions, completing the circuit. A schematic representation of a unit cell, illustrating the operation principle explained above is shown in **Figure 1**.

The PEMFCs require the improvement of their performance and the decreasing cost. A principal factor for their performance and its cost are the electrodes, where Pt catalysts are commonly used. So, it required the development of low-cost catalysts, mainly for the oxygen reduction reaction (ORR) at the cathode and hydrogen, methanol or ethanol oxidation reaction (EOR) at anode. The catalysts should reduce the Pt content increasing the rates of the corresponding reactions. Looking for better catalysts is an important challenge, which should be addressed with a comprehensive approach that involves the understanding of the reasons for which a material is an efficient catalyst for a given reaction; in this way, materials can be designed with a "scientific" base.

In this sense, the use of simulations to study catalysis is a great tool, particularly, with the use of atomistic models based on DFT and molecular dynamics. DFT is suited to study

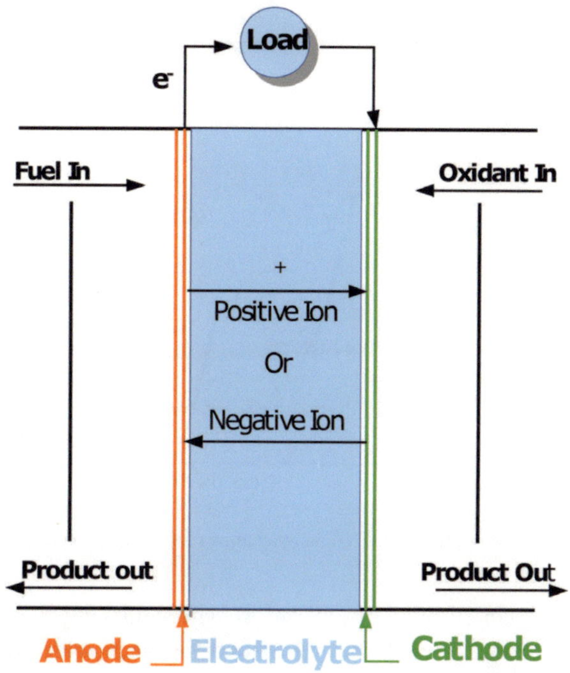

Figure 1. Schematic of a fuel cell.

adsorbate-catalyst interactions, which allow determining the potential energy surface and thus the energy barrier of the chemical transformations, also allows calculating the electronic structure of materials, which can help to understand bond-breaking and bond-making process. Also DFT, through thermodynamics approach, facilitates to describe reaction process and energetics of intermediates, reactants, and products under realistic conditions. Additionally, the information from atomistic scale can be used to compute reaction rates. In the next paragraphs, some works about catalysts for PEMFCs are presented.

Heterogeneous catalytic reactions in PEMFCs involve adsorption of reactants, different bond breaking and/or bond making processes and some electron transfer processes, and desorption of products. The reaction follows the path with minimal energy to go from reactants to product configuration, so this path depends, essentially, on the energetics of the possible elementary steps. So, many studies of catalysts applying DFT is focused in considering each possible reaction step to calculate energetics of reactants and products of each step from which energy barrier for each step can be calculated with different methods such as nudged elastic band [15–17], synchronous transit-guided quasi Newton (STQN) methods [18], among others. With this method, a free-energy diagram for different process can be drawn, an example for dissociative mechanism of oxygen on Pt is shown in **Figure 2**.

The above described approach was used to study EOR or the ethanol decomposition in other processes on Pt [19, 20], Rh [21], Pd [22], Au [23], and some Pt alloys as PtRh, PtRe, PtRu, PtSn, and PtPd [24]. In some of these works, only some steps or the complete reaction network was considered. For example, it was calculated by Alcalá et al., the transition states for C-C and C-O

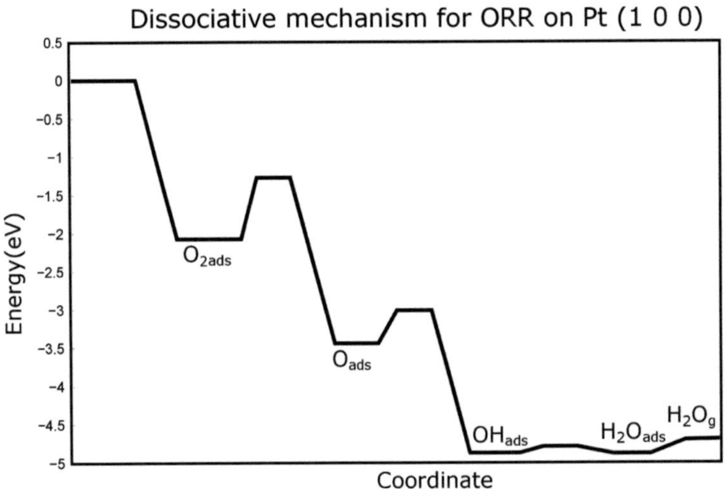

Figure 2. Potential energy surface diagram of dissociative mechanism for oxygen reduction reaction on Pt (1 0 0).

bond cleavage on Pt using constrained optimization, where the length of the bond representing the reaction coordinate is constrained; they identified the steps with lowest activation energy for these bond cleavage reactions. The evaluation of complete reaction network is a formidable task due to the higher quantity of possible intermediate derivatives, but in the ethanol oxidation, reaction network is explored thoroughly identifying key steps for the selectivity on ethanol oxidation and the intermediates, which can lead to C-C bond cleavage in some surfaces.

Also, the mechanistic details of ORR on Pt and Pt-based alloys were investigated using DFT to calculate the intermediate energies and to estimate reaction and barrier energy of each step considered in the reaction pathway [25–33]. Different pathways are considered such as the O_2 dissociation mechanism, and the associative mechanism implying the addition of H to the O_2-containing species. Also, it considered two different ways for hydrogen transfer reaction within each of these pathways: the direct addition of H from the electrolyte (Eley-Rideal type) or the H adsorption and subsequent transfer to O_2-containing species (Langmuir-Hinshelwood type) [27]. For example, in [30], the dissociative mechanism finding that the high stability of oxygen and hydroxyl can imply kinetic limitations in proton/electron transfer to adsorbed oxygen or hydroxyl was studied. In other study, different mechanisms were considered and it is found that under determined conditions, the determining step is the O_2 dissociation, so the associative mechanism is preferred [27].

One of the challenges in the study of electrochemical reactions is considering the effects of the solvation of molecules, the electric field in the interface electrolyte-electrode, and the other adsorbed molecules, nonetheless, some developments have successfully addressed those issues. It is widely accepted that the water layers between the polymer and the catalysts are very important in ORR. It was intended to study the effects of solvation using nine water molecules and one hydronium ion in an ab initio molecular dynamic study [34]. Also, in other studies, adsorption energies of intermediates were calculated adding water adsorbed to the surface or water on top of the adsorbed intermediates [30, 35]. The solvent can be also described in averaged manner as dielectric continuum characterized by a dielectric constant, which is used to estimate free energy of solute-solvent interactions [36–38]; this approach is named implicit solvation method.

As it is envisioned above, significant progress for the accurate description of electrochemical reactions have been achieved. However, it also can be noted that the complete kinetic description of a given system is quite a demanding task. But, different works show that enough knowledge about catalytic activity could be gained studying some properties of the system at the atomic scale, which can be calculated in a simpler way. These properties are named descriptors, that is, an observable that can be directly linked to a fundamental property [10, 14]; in this case, the catalytic activity. For example, Nørskov et al. show that the oxygen binding energy on metal electrodes is related to the turnover frequency of the oxygen reduction reaction [30].

ORR involves two key steps, the H transfer to adsorbed O_2 to form adsorbed OOH before the O-O bond breaking and the H transfer to adsorbed OH to form H_2O. The activation energy of each step is related with the energy change in each process, which in turn is related with stability of adsorbed OOH and OH, respectively. These adsorbed species will be more stable if O binding to the surface is stronger; therefore, the O binding energy can be related with the reaction rates of these two key steps. The reasoning for this relation is as follows: If O binding is stronger, it is difficult to destabilize the Pt-OH bond to form H_2O; if it is weak, it is difficult to stabilize the OOH to allow the O-O bond to break, so the O binding energy should be moderated. Precisely, in different works, bimetallic Pt and Pd alloys promising as catalysts for ORR looking for materials with moderated oxygen binding energy were identified [39, 40].

The descriptor approach also was used to search materials for hydrogen oxidation reaction on anode of PEMFC. For this reaction, the binding energy of H atom is a key, and according to Sabatier principle, optimal catalytic activity can be achieved on surfaces with moderated binding energies [9, 41]. Sabatier principle states that the bond between the catalyst and the key adsorbate should not be too weak, such that the reactants do not bind, and are not too strong, since this leads to catalyst poisoning [42]. Similarly, in other work, the CO binding energy is considered along the H binding energy to find active catalysts tolerant to CO poisoning [11].

Other property that can be used as a descriptor of catalytic activity is the d-band center of the transition metal surfaces, that is, the average energy of the d states. As is proposed by the d-band model [6, 43], the d-band center is correlated with the binding energy of adsorbates to the surface. The d-band model is fulfilled for bonding of several atoms and molecules, then, as is shown in several works, tuning the electronic structure of a metal by the addition of alloying atom can lead to a better catalyst for ORR [44–46] or catalysts more tolerant to CO [47]. Also, it identified other electronic structure factors, such as transition metal, e.g., band filling and covalency, that can be used as descriptors for ORR on metal oxide catalysts [13].

It is worth to mention that the review presented is very brief and there are other extensive reviews that should be consulted [48, 49]. Besides, from the works reviewed above, it should be remarked that it seems that one key property for catalytic activity is the adsorption energy of key intermediate molecules, which can be used to search better catalysts for a determined reaction. Next section reviews the principal concepts in catalysis used for the identification of appropriate descriptor, in this sense scaling relations between activation energies and binding energies, Sabatier principle and d-band model are reviewed.

3. Development of catalysts using chemical descriptors

In the last section, different properties of a material that can be related with its catalytic activity were observed. For instance, binding energies of atoms and molecules can be related with activation energies, so with reaction rates. Sabatier principle shows that a determined value of the adsorption energy of intermediates will lead to higher catalytic activity. Finally, it is mentioned that modification of the electronic structure changes the catalytic activity.

3.1. Correlations between adsorption and activation energies and Sabatier principle

The central issue in catalysis is the effect of the catalyst on the rate of a chemical reaction, the rate constant of an elementary reaction can be calculated using an Arrhenius expression in terms of a pre-factor (v) and an activation energy (E_a)as follows:

$$k = ve^{-E_a/k_B T} \tag{1}$$

The activation energy is an important quantity, which varies with the change of catalyst, so reviewing the origin of activation energy is useful to understand the effect of catalyst on the reaction rate.

As it is illustrated in **Figure 3**, any chemical reaction can be described as a transition between two local minima on the potential energy surface of the system as a function of the spatial coordinates of all the involved atoms. The reaction path is defined as the minimum energy path, which connects the configuration of the reactants with minimum energy (R) to the

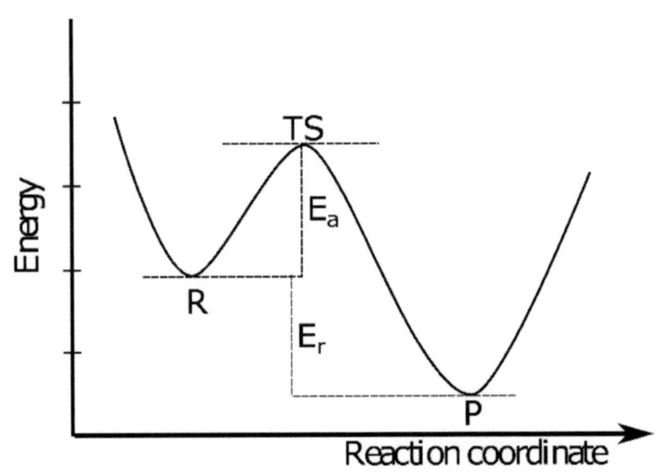

Figure 3. Potential energy surface of a chemical reaction process.

configuration of the products with minimum energy (P) along the reaction path; there is a configuration with the highest energy, which is called transition state (TS). The difference between the TS and R energies is the activation energy (E_a) [50–52]. The configurations of R, P, and TS in a catalytic process involve adsorbed molecules, then the energy of these configurations, and hence activation energy, depends on each surface.

$$E_a = E_{TS} - E_R \tag{2}$$

Also, it is worth to remark that as transition states have the same basic physics as adsorbed species, then the transition state energy (E_{TS}) is correlated with adsorption energies (E_{ads}). In fact, linear correlations between transition state and adsorption energies have been found [53–56]. This type of relationship is a well-established approach in the understanding of trends in chemical reactions that dates back to Brønsted in 1928, also Evans and Polanyi in 1938 [57], who stated a linear relationship between activation energies and reaction energies, named Brønsted-Evans-Polanyi (BEP) relationship.

The linear relationship between E_{TS} of a chemical reaction and a particular E_{ads} relevant for that reaction can be understood thinking that TS is an absorbed specie with a configuration relatively close to the adsorption configuration of the molecule X with which is correlated. Then, it is reasonable to express the E_{TS} as the sum of two contributions, the first is part of adsorption energy of the molecule X (αE_{ads-X}), which accounts, mainly, for the interaction energy between the surface and the molecule. The second is the quantity of energy (β), which accounts for the reorganization of the molecule X to transform into the transition state molecule. So, the general form of linear relationship is:

$$E_{TS} = \alpha E_{ads} + \beta \tag{3}$$

It is common to use E_{ads} as the energy of products configuration. In this case, parameter α is a value between 0 and 1, which indicates the position of the transition state along the reaction coordinate. If alpha is close to 1, the transition state has a configuration similar to the products (late transition state) [56], whereas β is a part of the energy involved in the reverse reaction (energy to transform products into transition state molecule) [56].

Using DFT calculations to calculate transition state energies and product energies, it has been possible to establish transition state scaling relations with sufficient statistical significance. This linear relationship provides a straightforward way to study the effect of the surface on the activity for a given reaction, as it enables the estimation of activation energies from adsorption energies, which could be calculated in a simpler way using computational calculations.

Transition state scaling relationship implies that the activation energies are lower in surfaces, where the products are more stable (more negative adsorption energies of products) and hence the reaction rate is higher. However, the very strong adsorption of products leads to poison the surface, since the desorption of products is more difficult, hence the rate of desorption is lower. On the other hand, if adsorption of products is very weak, the reaction rate would be lower and the desorption rate higher. Therefore, the adsorption energy of products

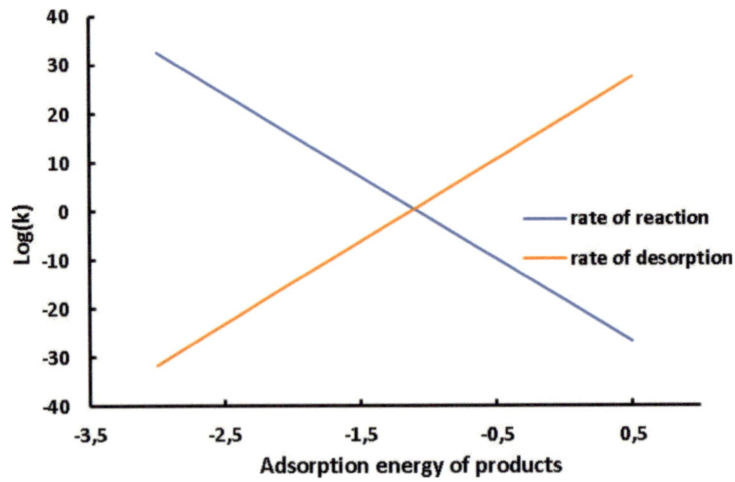

Figure 4. Volcano plot obtained from simple model.

should not be too weak, such that the reaction does not happens, and not too strong since this leads to catalyst poisoning [42]. The above mentioned is precisely the Sabatier principle, a valuable statement for catalyst design.

It allows an explanation of volcano curves, which show the catalytic activity for a given elementary reaction as function of the adsorption energy. As shown in **Figure 4**, at low values of adsorption energy (more positive values), the reaction is slow because the rate of the reaction is rate-limiting, whereas at high values of adsorption energy, desorption becomes the rate-limiting step. Then, the intermediate values of adsorption energies are required in order to obtain the highest activity (top of the volcano) [42, 58–60]. The volcano plot in the figure was obtained from a theoretical calculation, where the rate constants for temperature value of 298 K are obtained from Eq. (1) ($k = ve^{-E_a/k_BT}$), the activation energy for the reaction is obtained from Eqs. (2) and (3) ($E_a = E_{TS} - E_R$, $E_{TS} = \alpha E_{ads} + \beta$) with α and β equal to 1, and E_R equal to −1.2. The activation energy for desorption process is the negative of E_{ads}.

As mentioned above, a key issue in catalysis is chemisorption, the bond formed between surface and adsorbate can be described in terms of electronic structure of the surface. For transition metals, it is described in terms of d-bands of metals, as is proposed in the "d-band model," which is presented in the next subsection.

3.2. d-band model

The electronic structure of transition metals involves a broad s-band and narrow d-band. The interaction of the adsorbate valence electrons with s-band electrons can be assumed to be similar for all the different transition metals; hence, the principal differences in the interaction emerge from d-band electrons [4, 5, 43]. The interaction between the adsorbate and surface d-electrons can be represented with molecular orbitals theory, as illustrated in **Figure 5**, the interaction causes the formation of new shared states, which can be classified into bonding and antibonding states; the antibonding states are higher in energy and its filling leads to the weakening of the bond between adsorbate and surface [4, 5, 43, 61].

If more of antibonding states are empty, then the surface bonds strongly the molecule. Since electrons fill up all states located below the Fermi energy, in order to estimate the strength of the bond is enough to know how many antibonding states are higher in energy than Fermi energy (E_F). It has been found that if d-bands of clean surface shifts up in energy, the anti-bonding states also would shift up in energy and probably would be emptier. In this sense, the "orange surface" of **Figure 5** should bind the adsorbate stronger than blue surface, since it has d-states with higher energy and so the antibonding states would be above Fermi energy (emptier antibonding states).

The average is a representative value for a group the values or a statistical distribution is the average. In this case, energy average of d-states can be used to indicate the energy of all metal d-states, and it is calculated using Eq. (4) [51]. The difference between this value and Fermi energy can be used to compare in an effective way the d-states energy of different metals. This value is named d-band center (ε_d), if it is higher (less negative), then more antibonding states would be above Fermi energy and the bond would be stronger. This is the called d-band model proposed by Hammer and Nørskov [6, 43].

$$\varepsilon_d = \frac{\int_{-\infty}^{+\infty} n_{d(E)EdE}}{\int_{-\infty}^{+\infty} n_d(E)dE} \tag{4}$$

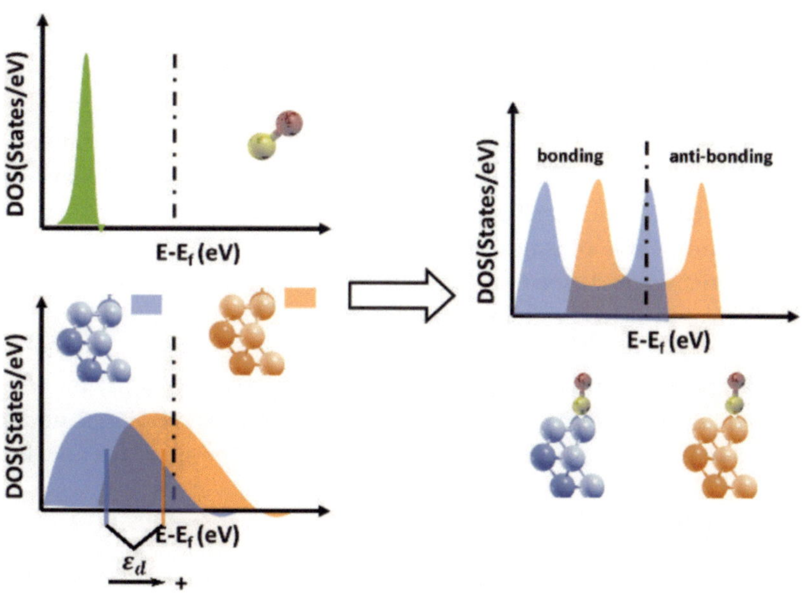

Figure 5. Illustration of the d-band model. The interaction of adsorbate states (green) with metal d-states (blue and orange) forms bonding and anti-bonding states. As the d-band center of metal shift down in energy (blue), more antibonding states are below E_F (dotted line) and hence bond is weaker.

The concepts above presented represent the basic theoretical instruments used to investigate catalytic chemical reactions and to provide satisfactory descriptions with a reduced set of properties compared to complete kinetic description.

3.3. Summarizing the descriptors approach

As remarked in the above sections, the descriptor approach is based on the linking of catalytic activity with a limited set of properties, named descriptors. The main step of this approach is to find a suitable reactivity descriptor and the other is to calculate it. As it was remarked, the adsorption energies of key intermediates can be used as descriptors, the identification of particular adsorption energies that are descriptors can be achieved considering the underlying reaction mechanism in order to identify the rate limiting step, also a Sabatier analysis and the known BEP relations can be used to identify the key adsorption energy. Also, an adsorption energy descriptor can be found through an educated guess, resulting in the knowledge on similar reactions [4, 5, 14].

Since adsorbate-surface interaction is a phenomenon originated by electronic interaction, the electronic structure of the systems can be related with adsorption energies, and some characteristics of this electronic structure can be used as descriptor, one of them is the d-band center. But additional research is required to identify more reliable properties, which can be used in other systems different to transition metals. The next section shows the main results of the modeling with chemical descriptors approach of catalytic carbon-carbon bond cleavage in the ethanol oxidation, reaction important for PEMFCs.

4. Analysis of C-C bond cleavage using chemical descriptors

The catalytic cleavage of the C-C bond in ethanol is fundamental for improving the proton exchange membrane fuel cells fueled with ethanol, whose efficiency is still far from obtained with hydrogen fuel cells, which can be partly attributed to the incomplete oxidation of ethanol to CO_2, so energy from ethanol is not totally exploited [62–67]. Catalysts actually used only promotes partial ethanol oxidation to acetic acid, the limiting step for total oxidation to CO_2 is, precisely, the C-C bond cleavage. Some works suggest that addition of metals as Ni [68], Rh [69], Ir [70], and Re [71] to Pt helps in C-C breakage at temperatures below 100°C. So, it is of interest to model this reaction step on Pt-based catalysts in order to clarify the effect of different co-catalysts on the C-C bond cleavage and to understand the conditions under which the C-C bond can be broken. It can be achieved using the descriptor approach above presented.

4.1. Reactivity descriptor for C-C bond cleavage

The C-C cleavage in molecules with more hydrogen atoms attached to the carbon atoms, as ethanol molecule (CH_3CH_2OH) or acetaldehyde (CH_3CHO), has higher activation energies [19, 21, 22]. It is attributed to the high directionality of C-C bond σ-orbital in these molecules, which is constrained along the bond axis, and the several substituents (hydrogen and oxygen atoms) on both carbon atoms constraining sterically this bond, which makes the interaction of the metal with this bond more difficult than with C-H single bond [72]. Probably, breaking of C-H bonds prior to C-C bond cleavage is required, thus the metal could interact with π-orbitals of double C-C bond, which are oriented sideway and have less steric constrains [72]. In fact, it is found that in hydrocarbon reaction on metal surfaces, molecules dehydrogenate before the cleaving of C-C bond [73].

Experimental works on ethanol oxidation reaction [62, 74–76] shows that ethanol oxidation involves dehydrogenation reactions, forming acetyl CH_3CO; this molecule could be further dehydrogenated forming CH_2CO and CHCO. In all the molecules, the C-C bond cleavage is more feasible, so it is worth to focus on them. The reaction for C-C bond cleavage in these CH_XCO molecules is written as follows

$$CH_x\,CO_{(ads)} \rightarrow CH_{x(ads)} + CO_{(ads)} \tag{5}$$

Linear relationship between products energy and transition state energy has been previously developed for this reaction [19, 22, 53, 77]. So, the activation energy would be related with the interaction of the surface with CH_X and CO molecules. It is expected that the trends in the reactivity do not changes with the molecule CH_XCO chosen, since the adsorption energy of the CH_X molecules is expected to be similar [53], that is, if rhodium has higher adsorption of CH than Ru, it would have higher adsorption energy of CH_2 and CH_3 than Ru. So, in this work, to study the C-C bond cleavage in the simplest molecule CHCO, also this step, according to previous works [19, 22], has the lowest activation energy for the C-C bond cleavage on different transition metals. The linear relationship [77], which will be used, is shown in Eq. (6).

$$E_{TS} = \alpha E_P + \beta \; with \; \alpha = 0.88 \; \beta = 1.07 \tag{6}$$

E_{TS} and E_P are, respectively, the transition state and product energies respect to the energy of reactants in gas phase and the clean slab. So, E_P is the energy involved to go from CHCO and the surface far apart to the CH and CO adsorbed on the surface that is expressed in Eq. (7)

$$E_P = E_{ads\text{-}CO} + E_{ads\text{-}CH} + E_{diss\text{-}CHCO} \tag{7}$$

Activation energy is expressed as in Eq. (2), in this case, E_R is the energy of $CHCO$ adsorbed respect to the energy of reactants (CHCO) in gas phase and the clean slab, that is, the adsorption energy of CHCO. So, the final expression for activation energy, which states the relationship between catalytic activity and adsorption energy, is:

$$E_a = \alpha(E_{ads\text{-}CO} + E_{ads\text{-}CH} + E_{diss\text{-}CHCO}) + \beta - E_{ads\text{-}CHCO} \tag{8}$$

Considering that CO_2, which desorbs from surface easily, can be readily obtained from CO oxidation by OH, which is available in alkaline media and in acidic media can be obtained from water activation facilitated by oxyphilic atoms as Sn, Ru, and Mo [20, 75, 78, 79]. So, it seems reasonable to expect that in the appropriated conditions (alkaline media or oxyphilic metal as co-catalyst), the surfaces with higher CO formation also have higher CO_2 formation.

Therefore, the study of the C-C bond cleavage step can give enough information about which catalysts can helps to obtain higher CO_2. The reactivity of the surfaces for C-C bond cleavage is studied through DFT modeling of adsorption of CH, CO, and CHCO.

The adsorption energies of these molecules were calculated for different Pt_3M alloys (M = Sn, Re, Ru, Rh, and Ni). Since the alloys with this composition or lower content of M are commonly proposed as a convenient composition to improve the catalytic activity to ethanol oxidation in experimental studies [68, 69, 71, 80–82].

4.2. Calculation of reactivity descriptors

The adsorption energies of CH, CO, and CHCO are calculated using the slab model to represent the (111) facet of the face-centered cubic metals (Pt_3M) and DFT as is implemented in Quantum Espresso [83]. It used plane waves, exchange correlation energy described by Perdew and Wang (PW91) [84], and ultra-soft pseudopotentials [85]. Convergence was tested with respect to total energy for the Brillouin sampling and for the kinetic energy cut-off in the plane wave expansion, so 3 × 3 × 1 Monkhorst-Pack k-mesh [86] and the kinetic energy cut-off for the plane waves of 544 eV was used.

For slab construction, the lattice parameters obtained from geometric optimization with DFT was used. The effect of the size of the supercell has also been carefully tested for the adsorption energy. Increasing the supercell size from 2 × 2, which we have used, to 4 × 4 changes the adsorption energy by only 0.008 eV, so the description with 2 × 2 is considered good enough considering the computational cost of the simulation of large systems. Five layers of atoms were employed for representing the portion of solid and empty space of 9.00 Å was left above of the atomic surface to avoid self-interaction of the slab. With this empty space, the adsorption energies changed just 0.02 eV compared to the adsorption energies obtained using an empty space of 15.00 Å, indicating that the top of one slab has essentially no effect on the bottom of the next.

The adsorption energy was calculated as total energy of the slab with the i-species adsorbed $E_{i/s}$ minus the total energy of clean slab E_s and the i-species in gas phase E_i.

$$E_{ads-i} = E_{i/s} - E_s - E_i \tag{9}$$

Adsorption of CH, CO, and $CHCO$ were tested on several sites, atop sites (A-Pt and A-M), bridge sites (B-Pt and B-M), and hollow sites with and without an atom directly below in the subsequent layer, which are called HCP and FCC hollow sites (HCP-Pt, HCP-M, FCC-Pt, and FCC-M).

The most stable adsorption configuration for CO adsorption on Pt_3M is at atop site with molecular axis of CO normal to the surface and C atom bonded to the surface in agreement with the experimental findings for Pt-based catalysts [87]. The CO adsorption energies on A-Pt and A-M for the different surfaces are shown in **Figure 6**. It is found that CO adsorption energy on A-Pt sites in the alloys is reduced respect to pure Pt (dotted line) and it is more reduced when the metal atom adsorbs more strongly the CO molecule. Also it is remarkable that adsorption energy on tin atom is only −0.08 eV. It is in agreement with other works, which points out that Pt-CO binding is weakened in the PtM alloy [88–90], whereas the M-CO bond is strengthened with the presence of Pt and CO adsorption on tin is not stable [91–93].

The most stable adsorption configuration for CH adsorption on Pt_3M is at hollow site (both hollow sites have similar adsorption energy) with molecular axis of CH normal to the surface and C atom bonded to an atom on the surface in agreement with other theoretical studies [94–96]. It is noted in **Figure 6**, a trend similar to CO adsorption, adsorption on sites with only Pt in the alloy is weaker than pure Pt, whereas the sites with metal atoms have higher or similar adsorption energy than pure Pt with the exception of Pt_3Sn, which is significantly lower.

The most stable adsorption configuration for CHCO on the alloys (except for Pt_3Sn) and pure Pt is shown in **Figure 7**. In this geometry, the CH part is located at bridge position between M and Pt atom (B-M), whereas the CO part is located close to a Pt atom in a Pt-atop-like position (A-Pt). For both groups, the C atom is closer to the surface. This configuration is like other found in theoretical studies for transition metals [77, 97]. Other adsorption configuration is found when CHCO is close to hollow-Pt sites, except for Pt_3Rh, whose Pt sites have an adsorption configuration as the above described. In this configuration, the carbon atom linked to the hydrogen atom is bound to the surface in a bridge position between two Pt atoms (B-Pt), whereas the carbon atom linked to oxygen is far from the surface as is shown in **Figure 7**. This configuration was found in other work for metals considered inactive [97].

Adsorption energies of former configuration with CH and CO parts bonded to the surface ranges from −3.66 to −3.98 and C-C bond length is stretched by 0.20 ± 0.01 Å, whereas adsorption energies of last configuration ranges from −2.68 to −3.24 eV and C-C bond length is only

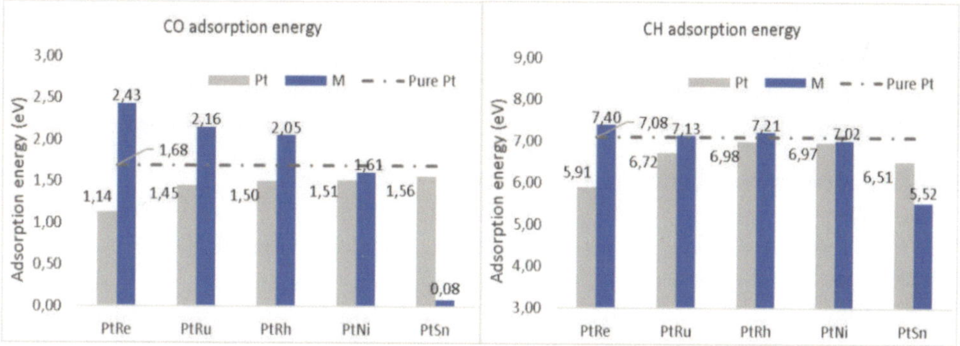

Figure 6. CO and CH and CHCO adsorption energies on Pt(111) and $Pt_3M(111)$ surfaces.

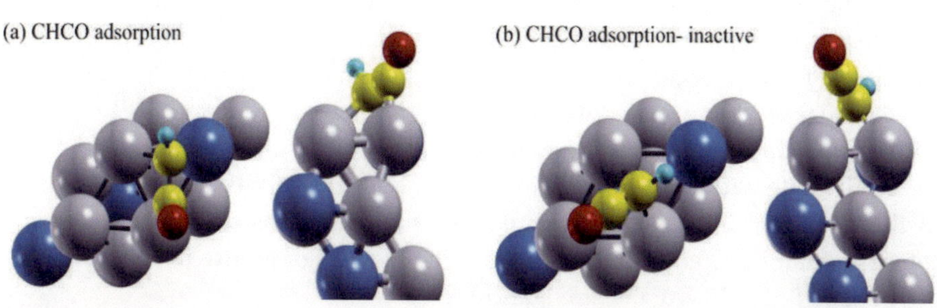

Figure 7. Geometric configurations found for CHCO adsorption on $Pt_3M(111)$ surfaces.

stretched slightly by 0.08 ± 0.01 Å. So, the ensembles with only Pt atoms in Pt_3M (with exception of Pt_3Rh) have adsorption energies significantly lower than pure Pt and do not stretch significantly the C-C bond, whereas the ensembles with M atom have adsorption energies similar to Pt and stretch significantly the C-C bond. Indicating that the addition of M to Pt decreases the ability of Pt for interacting with CHCO and hence part of the alloy is inactive, except for Pt_3Rh.

Summarizing, the adsorption of CH, CO, and CHCO shows that the addition of M (M = Re, Rh, Ni, Ru, or Sn) to Pt leads to obtain two active sites with different reactivities. The Pt atoms in all these alloys form bonds with the described molecules weaker than Pt atoms in pure Pt, whereas the Re, Rh, and Ru atoms in the alloys have a stronger interaction with CH, CO, and CHCO than pure Pt. This difference explains that adsorption energies of CO on metal atoms in Pt_3M (M = Rh, Re, and Ru) are higher than Pt (111), whereas the adsorption energies of CH on sites with M atoms in Pt_3M (M = Rh, Re, and Ru) are similar to Pt (111), since CH interacts with two Pt and one M atom at hollow site. Then, the stronger interaction of mentioned M atoms should compensate the weaker interaction of Pt in these alloys.

Besides, Ni and Pt atoms in Pt_3Ni have slightly weaker interaction than pure Pt (111), so the adsorption energies at different sites in this alloy are slightly lower than pure Pt. On the other hand, Sn and Pt atoms in Pt_3Sn have a weaker interaction with the molecules than pure Pt (111), suggesting that Pt_3Sn has low catalytic activity toward C-C bond cleavage, confirming previous experimental and theoretical works that affirms that PtSn alloys have lower catalytic activity for C-C bond cleavage, but they increase the catalytic activity for EOR by promoting bifunctional mechanism and the pathway toward acetic acid [70, 98]. Other remarkable finding is the two configurations for CHCO adsorption, one of them destabilizes the molecule binding it strongly and stretching significantly the C-C bond (around to 0.20 Å) and the other is found in metals inactive, it suggests that part of alloy is inactive, except for Pt_3Rh.

The adsorption energies obtained can be related with activation energy for C-C bond cleavage through Eq. (8), only the active sites are considered since the linear relationship between ETS and Eads only have sense for them. As is can be expected from Eq. (8), the surfaces with higher adsorption energies of CH and CO have lower activation energy, that is the sites with Re, Rh, and Ru atoms, it can be observed in **Figure 8**, which shows the product energies E_p (see Eq. (7)) and activation energies for the different active sites.

These results indicate that Pt_3Sn is the catalyst with the lowest activity for C-C bond cleavage, since in this alloy Sn has a weak interaction with all molecules studied and Pt atoms have a weaker interaction than Pt atoms in pure Pt (111). Also, it is expected that Pt_3Rh alloy has higher reactivity for C-C bond cleavage since all sites, sites with only Pt, and sites with Rh atoms have similar or higher reactivity than pure Pt. Whereas Pt_3Re and Pt_3Ru have a nonsignificant improvement of the reactivity with respect to pure Pt, since they have one site with higher reactivity (site with M atom) and other with lesser reactivity (site with only Pt atoms); between these ones, Pt_3Re has the most reactive site. Pt_3Ni has slightly weaker interaction with molecules than Pt. Consequently, the ranking from high to low reactivity for C-C bond cleavage is Pt_3Rh, Pt_3Re, Pt_3Ru, Pt, Pt_3Ni, and Pt_3Sn.

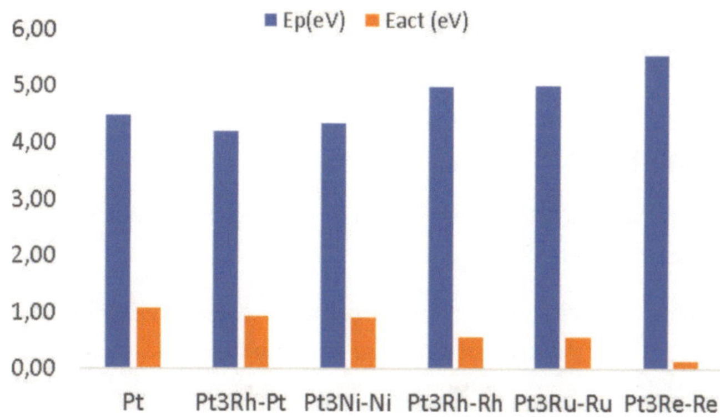

Figure 8. Final state energy (E_p) relative to reactants (CHCO) in gas phase and clean slab and activation energy on the different surfaces sites found active.

This trend agrees with previous experimental results, where it is reported that Pt-based alloys with rhodium (PtRh and PtSnRh) show a production of CO_2 higher than Pt and Pt alloys [69, 99]. In contrast, addition of Rhenium to Pt improves performance for C-C cleavage, but this improvement is not much higher, it is reported that Re addition to Pt improves the acetaldehyde decomposition, increasing the produced CO (product of C-C bond cleavage) from 10 to 18% [100]. On the contrary, it is reported that Sn addition to Pt decreases the CO_2 product respect to Pt [70, 98]. But the PtSn alloys are recognized as the better catalyst for EOR, where the enhancement of the activity with this alloy is attributed to the promotion of bifunctional mechanism promoting the acetic acid production.

Due to mentioned issues, different authors propose materials with Pt, Sn, and Rh as effective catalysts in splitting the C-C bond and oxidizing intermediates molecules of EOR via bifunctional mechanism. This mix in the appropriate composition will provide high overall conversion of ethanol and higher selectivity to CO_2 than Pt. However, the selectivity can be considered low and the total conversion to CO_2 remains elusive. It can be attributed to the presence of oxidants as OH, which should promote the oxidation of intermediates to acetic acid and decreases the C-C bond cleavage rate [101]. But the unavailability of oxidants limits the conversion of CO to CO_2, poisoning the surface. Finding a better catalyst for EOR requires considering, in addition of C-C cleavage, the step of oxidation to produce acetic acid and the dehydrogenation step.

Author details

Alejandro E. Pérez and Rafael Ribadeneira*

*Address all correspondence to: reribade@unal.edu.co

Department of Processes and Energy, Facultad de Minas, Universidad Nacional de Colombia-Sede Medellín, Colombia

References

[1] Kaxiras E. Atomic and Electronic Structure of Solids. New York, United States of America, Cambridge: Cambridge University Press; 2003

[2] Callister WD, Rethwisch DG. Materials Science and Engineering: An Introduction. 7th ed. John Wiley & Sons, Inc; 2007

[3] Seh ZW, Kibsgaard J, Dickens CF, Chorkendorff I, Nørskov JK, Jaramillo TF. Combining theory and experiment in electrocatalysis: Insights into materials design. Science. 2017; **355**(6321):eaad4998

[4] Nørskov JK, Bligaard T, Rossmeisl J, Christensen CH. Towards the computational design of solid catalysts. Nature Chemistry. 2009;**1**(1):37-46

[5] Nørskov JK, Abild-Pedersen F, Studt F, Bligaard T. Density functional theory in surface chemistry and catalysis. Proceedings of the National Academy of Sciences of the United States of America. 2011;**108**(3):937-943

[6] Hammer B, Nørskov JK. Electronic factors determining the reactivity of metal surfaces. Surface Science. 1995;**343**(3):211-220

[7] Linic S, Jankowiak J, Barteau MA. Selectivity driven design of bimetallic ethylene epoxidation catalysts from first principles. Journal of Catalysis. 2004;**224**(2):489-493

[8] Greeley J, Mavrikakis M. Alloy catalysts designed from first principles. Nature Materials. 2004;**3**(11):810-815

[9] Greeley J, Jaramillo TF, Bonde J, Chorkendorff I, Nørskov JK. Computational high-throughput screening of electrocatalytic materials for hydrogen evolution. Nature Materials. 2006;**5**(11):909-913

[10] Andersson M, Bligaard T, Kustov A, Larsen K, Greeley J, Johannessen T, et al. Toward computational screening in heterogeneous catalysis: Pareto-optimal methanation catalysts. Journal of Catalysis. 2006;**239**(2):501-506

[11] Strasser P, Fan Q, Devenney M, Weinberg WH, Liu P, Nørskov JK. High throughput experimental and theoretical predictive screening of materials–A comparative study of search strategies for new fuel cell anode catalysts. The Journal of Physical Chemistry. B. 2003;**107**(40):11013-11021

[12] Grabow LC, Hvolbæk B, Nørskov JK. Understanding trends in catalytic activity: The effect of adsorbate-adsorbate interactions for CO oxidation over transition metals. Topics in Catalysis. 2010;**53**(5-6):298-310

[13] Suntivich J, Gasteiger HA, Yabuuchi N, Nakanishi H, Goodenough JB, Shao-Horn Y. Design principles for oxygen-reduction activity on perovskite oxide catalysts for fuel cells and metal–air batteries. Nature Chemistry. 2011;**3**(8):647-647

[14] Grabow LC. Computational catalyst screening. In: Asthagiri A, Janik MJ, editors. Computational Catalysis. Cambridge, UK: Royal Society of Chemistry; 2013. pp. 1-58

[15] Jónsson H, Mills G, Jacobsen KW. Nudged elastic band method for finding minimum energy paths of transitions. In: Classical and Quantum Dynamics in Condensed Phase Simulations. Singapore: World Scientific; 1998. pp. 385-404

[16] Henkelman G, Uberuaga BP, Jónsson H. A climbing image nudged elastic band method for finding saddle points and minimum energy paths. The Journal of Chemical Physics. 2000;**113**(22):9901

[17] Henkelman G, Jónsson H. Improved tangent estimate in the nudged elastic band method for finding minimum energy paths and saddle points. The Journal of Chemical Physics. 2000;**113**(22):9978-9985

[18] Peng C, Bernhard Schlegel H. Combining synchronous transit and quasi-Newton methods to find transition states. Israel Journal of Chemistry. 1993;**33**(4):449-454

[19] Alcalá R, Mavrikakis M, Dumesic JA. DFT studies for cleavage of C-C and C-O bonds in surface species derived from ethanol on Pt(111). Journal of Catalysis. 2003;**218**:178-190

[20] Wang H, Liu Z. Comprehensive mechanism and structure-sensitivity of ethanol oxidation on platinum: New transition-state searching method for resolving the complex reaction. Network. 2008;**10**:10996-11004

[21] Choi Y, Liu P. Understanding of ethanol decomposition on Rh(111) from density functional theory and kinetic Monte Carlo simulations. Catalysis Today. 2011;**165**(1):64-70

[22] Li M, Guo W, Jiang R, Zhao L, Shan H. Decomposition of ethanol on Pd(111): A density functional theory study. Langmuir. 2010;**26**(3):1879-1888

[23] Ishimoto T, Kazuno H, Kishida T, Koyama M. Theoretical study on oxidation reaction mechanism on Au catalyst in direct alkaline fuel cell. Solid State Ionics. 2014;**262**:328-331

[24] Xu Z-F, Wang Y. Effects of alloyed metal on the catalysis activity of Pt for ethanol partial oxidation: Adsorption and dehydrogenation on Pt(3)M (M=Pt, Ru, Sn, Re, Rh, and Pd). The Journal of Physical Chemistry. C, Nanomaterials and Interfaces. 2011;**115**(42): 20565-20571

[25] Pérez-Mendoza AE, Castañeda-Ramírez S, Ribadeneira RE, Perez Mendoza AE, Castaneda Ramirez S, Ribadeneira Paz RE. Multiscale modeling of a proton exchange membrane fuel cell: Atomistic oxygen reduction reaction model. ECS Transactions. 2015; **66**(28):7-18

[26] Sidik RA, Anderson AB. Density functional theory study of O2 electroreduction when bonded to a Pt dual site. Journal of Electroanalytical Chemistry. 2002;**528**(1-2):69-76

[27] JA K, Jerkiewicz G, Jacob T. Theoretical investigations of the oxygen reduction reaction on Pt(111). Chemphyschem. 2010;**11**(13):2779-2794

[28] Eichler A, Mittendorfer F, Hafner J. Precursor-mediated adsorption of oxygen on the (111) surfaces of platinum-group metals. Physical Review B. 2000;**62**(7):4744-4755

[29] Eichler A, Hafner J. Molecular precursors in the dissociative adsorption of O2 on Pt(111). Physical Review Letters. 1997;**79**(22):4481-4484

[30] Nørskov JK, Rossmeisl J, Logadottir A, Lindqvist L, Lyngby D, Jo H. Origin of the overpotential for oxygen reduction at a fuel-cell cathode. The Journal of Physical Chemistry. B. 2004:**108**;17886-17892

[31] Walch S, Dhanda A, Aryanpour M, Pitsch H. Mechanism of molecular oxygen reduction at the cathode of a PEM fuel cell: Non-electrochemical reactions on catalytic Pt particles. Journal of Physical Chemistry C. 2008;**112**(22):8464-8475

[32] Balbuena PB, Altomare D, Agapito L, Seminario JM. Theoretical analysis of oxygen adsorption on Pt-based clusters alloyed with Co, Ni, or Cr embedded in a Pt matrix. The Journal of Physical Chemistry. B. 2003;**107**(49):13671-13680

[33] Panchenko A, Koper MTM, Shubina TE, Mitchell SJ, Roduner E. Ab initio calculations of intermediates of oxygen reduction on low-index platinum surfaces. Journal of the Electrochemical Society. 2004;**151**(12):A2016

[34] Jinnouchi R, Okazaki K. New insight into microscale transport phenomena in pefc by quantum md. Microscale Thermophysical Engineering. 2003;**7**(1):15-31

[35] Liu S, White MG, Liu P. Mechanism of oxygen reduction reaction on Pt(111) in alkaline solution: Importance of chemisorbed water on surface. Journal of Physical Chemistry C. 2016;**120**(28):15288-15298

[36] Sha Y, Yu TH, Liu Y, Merinov BV, Goddard WA. Theoretical study of solvent effects on the platinum-catalyzed oxygen reduction reaction. Journal of Physical Chemistry Letters. 2010;**1**(5):856-861

[37] Jinnouchi R, Kodama K, Hatanaka T, Morimoto Y. First principles based mean field model for oxygen reduction reaction. Physical Chemistry Chemical Physics. 2011;**13**(47):21070

[38] Mathew K, Sundararaman R, Letchworth-Weaver K, Arias TA, Hennig RG. Implicit solvation model for density-functional study of nanocrystal surfaces and reaction pathways. The Journal of Chemical Physics. 2014;**140**(8):084106-1-084106-8

[39] Greeley J, Stephens IEL, Bondarenko AS, Johansson TP, Hansen HA, Jaramillo TF, et al. Alloys of platinum and early transition metals as oxygen reduction electrocatalysts. Nature Chemistry. 2009;**1**(7):552-556

[40] Ou L. Design of Pd-based bimetallic catalysts for ORR: A DFT calculation study. Journal of Chemistry. 2015;**2015**:1-11

[41] Greeley J, Mavrikakis M. Near-surface alloys for hydrogen fuel cell applications. Catalysis Today. 2006;**111**(1-2):52-58

[42] Knozinger H, Deutschmann O. Heterogeneous catalysis and solid catalysts. In: Ullmann's Encyclopedia of Industrial Chemistry. Weinheim, Germany: Wiley-VCH Verlag GmbH & Co. KGaA; 2009

[43] Hammer B, Nørskov JK. Theoretical surface science and catalysis: Calculations and concepts. Advances in Catalysis. 2000;**45**

[44] Stamenkovic V, Mun BS, Mayrhofer KJJ, Ross PN, Markovic NM, Rossmeisl J, et al. Changing the activity of electrocatalysts for oxygen reduction by tuning the surface electronic structure. Angewandte Chemie, International Edition. 2006;**45**(18):2897-2901

[45] Stamenkovic VR, Mun BS, Arenz M, Mayrhofer KJJ, Lucas CA, Wang G, et al. Trends in electrocatalysis on extended and nanoscale Pt-bimetallic alloy surfaces. Nature Materials. 2007;**6**(3):241-247

[46] Bandarenka AS, Koper MTM. Structural and electronic effects in heterogeneous electrocatalysis: Toward a rational design of electrocatalysts. Journal of Catalysis. 2013;**308**:11-24

[47] Bligaard T, Nørskov JK. Ligand effects in heterogeneous catalysis and electrochemistry. Electrochimica Acta. 2007;**52**(18):5512-5516

[48] Lile JRD, Zhou S. Theoretical modeling of the PEMFC catalyst layer: A review of atomistic methods. Electrochimica Acta. 2015;**177**:4-20

[49] Shi Z. Application of first principles methods in the study of fuel cell air-cathode electrocatalysis. In: Zhang J, editor. PEM Fuel Cell Electrocatalysts and Catalyst Layers. London: Springer; 2008. p. 289-329

[50] Groß A, Gross A. Theoretical Surface Science. A Microscopic Perspective. Berlin, Heidelberg: Springer Berlin Heidelberg; 2009

[51] Norskov JK, Studt F, Abild-Pedersen F, Bligaard T. Fundamental Concepts in Heterogeneous Catalysis. Hoboken, United States of America: John Wiley & Sons, Inc; 2014. 208 p

[52] Eyring H. The theory of absolute reaction rates. Transactions of the Faraday Society. 1938;**34**:41-48

[53] Wang S, Temel B, Shen J, Jones G, Grabow LC, Studt F, et al. Universal Brønsted-Evans-Polanyi relations for C–C, C–O, C–N, N–O, N–N, and O–O dissociation reactions. Catalysis Letters. 2011;**141**(3):370-373

[54] Wang S, Vorotnikov V, Sutton JE, Vlachos DG. Brønsted–Evans–Polanyi and transition state scaling relations of furan derivatives on Pd(111) and their relation to those of small molecules. ACS Catalysis. 2014;**4**(2):604-612

[55] Nørskov JK. Universality in heterogeneous catalysis. Journal of Catalysis. 2002;**209**(2): 275-278

[56] Michaelides A, Liu ZP, Zhang CJ, Alavi A, King DA, Hu P. Identification of general linear relationships between activation energies and enthalpy changes for dissociation reactions at surfaces. Journal of the American Chemical Society. 2003;**125**(13):3704-3705

[57] Evans MG, Polanyi M. Inertia and driving force of chemical reactions. Transactions of the Faraday Society. 1938;**34**:11

[58] Bligaard T, Nørskov JKK, Dahl S, Matthiesen J, Christensen CHH, Sehested J. The Brønsted-Evans-Polanyi relation and the volcano curve in heterogeneous catalysis. Journal of Catalysis. 2004;**224**(1):206-217

[59] Nørskov JK, Bligaard T, Hvolbaek B, Abild-Pedersen F, Chorkendorff I, Christensen CH. The nature of the active site in heterogeneous metal catalysis. Chemical Society Reviews. 2008;**37**(10):2163-2171

[60] Logadottir A, Rod TH, Nørskov JK, Hammer B, Dahl S, Jacobsen CJH. The Brønsted–Evans–Polanyi relation and the volcano plot for ammonia synthesis over transition metal catalysts. Journal of Catalysis. 2001;**197**(2):229-231

[61] Nilsson A, Pettersson LGM, Hammer B, Bligaard T, Christensen CH, Norskov JK. The electronic structure effect in heterogeneous catalysis. Catalysis Letters. 2005;**100**(3-4):111-114

[62] García-rodríguez S, Herranz T, Rojas S. New and future developments in catalysis. In: Electroctalysts for the Electrooxidation of Ethanol. Amsterdam, The Netherlands: Elsevier; 2013. pp. 33-67

[63] Lamy C. The direct ethanol fuel cell: A challenge to convert bioethanol cleanly into electric energy. In: Barbaro P, Bianchini C, editors. Catalysis for Sustainable Energy Production. Weinheim, Germany: WILEY-VCH Verlag GmbH & Co. KGaA; 2009. pp. 1-46

[64] Kutz RB, Braunschweig B, Mukherjee P, Behrens RL, Dlott DD, Wieckowski A. Reaction pathways of ethanol electrooxidation on polycrystalline platinum catalysts in acidic electrolytes. Journal of Catalysis. 2011;**278**(2):181-188

[65] Badwal SPS, Giddey S, Kulkarni A, Goel J, Basu S. Direct ethanol fuel cells for transport and stationary applications – A comprehensive review. Applied EnergyAppl Energy. 2015;**145**:80-103

[66] Zhao TS, Li YS, Shen SY. Anion-exchange membrane direct ethanol fuel cells: Status and perspective. Frontiers of Energy and Power Engineering in China. 2010;**4**(4):443-458

[67] Kamarudin MZF, Kamarudin SK, Masdar MS, Daud WRW. Review: Direct ethanol fuel cells. International Journal of Hydrogen Energy. 2012;**38**:1-16

[68] Beyhan S, Coutanceau C, Léger JM, Napporn TW, Kadirgan F. Promising anode candidates for direct ethanol fuel cell: Carbon supported PtSn-based trimetallic catalysts prepared by Bönnemann method. International Journal of Hydrogen Energy. 2013;**38**:6830-6841

[69] Kowal A, Li M, Shao M, Sasaki K, Vukmirovic MB, Zhang J, et al. Ternary Pt/Rh/SnO_2 electrocatalysts for oxidizing ethanol to CO_2. Nature Materials. 2009;**8**(4):325-330

[70] Li M, Cullen DA, Sasaki K, Marinkovic NS, More K, Adzic RR. Ternary electrocatalysts for oxidizing ethanol to carbon dioxide: Making ir capable of splitting C-C bond. Journal of the American Chemical Society. 2013;**135**(1):132-141

[71] Tayal J, Rawat B, Basu S. Effect of addition of rhenium to Pt-based anode catalysts in electro-oxidation of ethanol in direct ethanol PEM fuel cell. International Journal of Hydrogen Energy. 2012;**37**(5):4597-4605

[72] Murakami M, Ito Y. Cleavage of carbon—carbon single bonds by transition metals. In: Activation of Unreactive Bonds and Organic Synthesis. Berlin Heidelberg: Springer; 1999

[73] Mueller JE, Van Duin ACT, Goddard WA. Application of the ReaxFF reactive force field to reactive dynamics of hydrocarbon chemisorption and decomposition. Journal of Physical Chemistry C. 2010;**114**(12):5675-5685

[74] Shao MH, Adzic RR. Electrooxidation of ethanol on a Pt electrode in acid solutions: In situ ATR-SEIRAS study. Electrochimica Acta. 2005;**50**(12):2415-2422

[75] Vigier F, Coutanceau C, Hahn F, Belgsir EMM, Lamy C. On the mechanism of ethanol electro-oxidation on Pt and PtSn catalysts: Electrochemical and in situ IR reflectance spectroscopy studies. Journal of Electroanalytical Chemistry. 2004;**563**(1):81-89

[76] Shen SY, Zhao TS, Wu QX. Product analysis of the ethanol oxidation reaction on palladium-based catalysts in an anion-exchange membrane fuel cell environment. International Journal of Hydrogen Energy. 2012;**37**(1):575-582

[77] Ferrin P, Simonetti D, Kandoi S, Kunkes E, Dumesic JA, Nørskov JK, et al. Modeling ethanol decomposition on transition metals: A combined application of scaling and Brønsted-Evans-Polanyi relations. Journal of the American Chemical Society. 2009;**131**(16): 5809-5815

[78] Antolini E, Gonzalez ERR. Effect of synthesis method and structural characteristics of Pt-Sn fuel cell catalysts on the electro-oxidation of CH_3OH and CH_3CH_2OH in acid medium. Catalysis Today. 2011;**160**(1):28-38

[79] Gong X-Q, Hu P, Raval R. The catalytic role of water in CO oxidation. The Journal of Chemical Physics. 2003;**119**(12):6324-6334

[80] Guo Y, Zheng Y, Huang M. Enhanced activity of PtSn/C anodic electrocatalyst prepared by formic acid reduction for direct ethanol fuel cells. Electrochimica Acta. 2008;**53**: 3102-3108

[81] Li H, Sun G, Cao L, Jiang L, Xin Q. Comparison of different promotion effect of PtRu/C and PtSn/C electrocatalysts for ethanol electro-oxidation. Electrochimica Acta. 2007;**52**(24):6622-6629

[82] Vigier F, Coutanceau C, Perrard A, Belgsir EMM, Lamy C. Development of anode catalysts for a direct ethanol fuel cell. Journal of Applied Electrochemistry. 2004;**34**(4):439-446

[83] Giannozzi P, Baroni S. QUANTUM ESPRESSO: A modular and open-source software project for quantum simulations of materials. Journal of Physics: Condensed Matter. 2009;**36**:395502-1-395502-36

[84] Perdew JP. Unified theory of exchange and correlation beyond the local density approximation. In: Ziesche P, Eschrig H, editors. Electronic Structure of Solids '91. Berlin: Akademie Verlag; 1991. pp. 11-20

[85] Vanderbilt D. Soft self-consistent pseudopotentials in a generalized eigenvalue formalism. Physical Review B. 1990;**41**(11):7892-7895

[86] Pack JD, Monkhorst HJ. "Special points for Brillouin-zone integrations"-a reply. Physical Review B. 1977;**16**(4):1748-1749

[87] Van Hove MA. The adsorption sites of CO and NO molecules on metal surfaces. Israel Journal of Chemistry. 1998;**38**(4):349-352

[88] Liao MS, Cabrera CR, Ishikawa Y. Theoretical study of CO adsorption on Pt, Ru and Pt-M (M = Ru, Sn, Ge) clusters. Surface Science. 2000;**445**(2-3):267-282

[89] Su HY, Bao XH, Li WX. Modulating the reactivity of Ni-containing Pt(111)-skin catalysts by density functional theory calculations. The Journal of Chemical Physics. 2008; **128**(19):1-7

[90] Ramstad A, Strisland F, Raaen S, Borg A, Berg C. CO and O_2 adsorption on the Re/Pt(111) surface studied by photoemission and thermal desorption. Surface Science. 1999;**440**(1-2):290-300

[91] Shubina TE, Koper MTM. Quantum-chemical calculations of CO and OH interacting with bimetallic surfaces. Electrochimica Acta. 2002;**47**(22-23):3621-3628

[92] Dupont C, Jugnet Y, Loffreda D. Theoretical evidence of PtSn alloy efficiency for CO oxidation. Journal of the American Chemical Society. 2006;**128**(28):9129-9136

[93] Liu P, Logadottir A, Nørskov JK. Modeling the electro-oxidation of CO and H_2/CO on Pt, Ru, PtRu and Pt3Sn. Electrochimica Acta. 2003;**48**(25-26):3731-3742

[94] Kua J, Faglioni F, W a G. Thermochemistry for hydrocarbon intermediates chemisorbed on metal surfaces: $CHn\_m(CH_3)_m$ with $n = 1, 2, 3$ on Pt, Ir, Os, Pd, Rh, and Ru. Journal of the American Chemical Society. 2000;**122**(5):2309-2321

[95] Hahn K, Mavrikakis M. Atomic and molecular adsorption on Re(0001). Topics in Catalysis. 2014;**57**(1-4):54-68

[96] Zhang M, Yang K, Zhang X, Yu Y. Effect of Ni(111) surface alloying by Pt on partial oxidation of methane to syngas: A DFT study. Surface Science. 2014;**630**:236-243

[97] Courtois J, Du W, Wong E, Teng X, Deskins NA. Screening iridium-based bimetallic alloys as catalysts for direct ethanol fuel cells. Applied Catalysis A: General. 2014;**483**: 85-96

[98] Alcala R, Shabaker JW, Huber GW, MA S-C, JA D. Experimental and DFT studies of the conversion of ethanol and acetic acid on PtSn-based catalysts. The Journal of Physical Chemistry. B. 2005;**109**(6):2074-2085

[99] Bach Delpeuch A, Maillard F, Chatenet M, Soudant P, Cremers C. Ethanol oxidation reaction (EOR) investigation on Pt/C, Rh/C, and Pt-based bi- and tri-metallic electrocatalysts: A DEMS and in situ FTIR study. Applied Catalysis B: Environmental. 2016; **181**:672-680

[100] Ciftci A, Ligthart DAJM, Sen AO, van Hoof AJF, Friedrich H, Hensen EJM. Pt-Re synergy in aqueous-phase reforming of glycerol and the water-gas shift reaction. Journal of Catalysis. 2014;**311**:88-101

[101] Kavanagh R, Cao XM, Lin WF, Hardacre C, Hu P. Origin of low CO_2 selectivity on platinum in the direct ethanol fuel cell. Angewandte Chemie, International Edition. 2012;**51**(7):1572-1575

Density Functional Theory Studies of Catalytic Sites in Metal-Organic Frameworks

Siwarut Siwaipram, Sarawoot Impeng,
Philippe A. Bopp and Sareeya Bureekaew

Additional information is available at the end of the chapter

http://dx.doi.org/10.5772/intechopen.80698

Abstract

Theoretical methods have become indispensable tools in many fields of chemistry and materials research. Metal-organic frameworks (MOFs) are porous materials; they have been intensively developed due to their diverse properties suitable for a wide range of applications. Theoretical approaches have thus been frequently employed toward the design and characterization of MOFs. We focus here in particular on theoretical studies of single-site catalytic reactions that occur inside the cavities of MOFs. The density functional method (DFT) has been the main approach used for such studies. We briefly review the uses of DFT to examine the catalytic reactions in MOFs. We note that DFT methods are versatile and can be made to work for different purposes such as, e.g., force-field development for molecular simulations. We shall, however, cover this field only very succinctly to put it into context with our main topic.

Keywords: theoretical chemistry, computational chemistry, quantum chemistry, density functional theory (DFT), metal-organic frameworks (MOF), catalysis, single-site catalytic reactions

1. Introduction

Experiments in chemistry aim primarily at understanding, or insight. These insights lead in turn to "innovation" and new technologies [1]. However, in many instances, such an insight cannot be obtained purely from current experiments. The required fundamental theory has been available for some time. The more recent development of computational methodologies and the dramatic increase in computer power made computational methods an additional and competitive tool to describe chemical phenomena [2]. This field is broadly known as "modeling." In "molecular modeling," the premise is that the interatomic and intermolecular interactions can be known [3]. DFT has become the dominant tool for the purpose of determining such interactions [4]. In particular, in solid-state chemistry and physics, it has become increasingly popular since appropriate functionals and basis sets [5] were designed to provide a viable balance between the reliability of the numerical results and the computational costs.

Metal-organic frameworks (MOFs), also known as porous coordination polymers (PCPs) [6–8], are a class of hybrid organic-inorganic materials with high surface area, a permanent porosity with large internal pore volume and tunable pore sizes. They have a wide range of potential applications such as gas storage [9] and separation [10], catalysis [11], and many others [12–14]. MOFs are constructed by interlinking metal ions or, more generally, metal-containing units with organic moieties (carboxylates, azolates, imidazolates, pyridyl, etc.) through coordination bonds, thus creating crystalline frameworks. Due to the variety of the structural and chemical elements, some MOFs show unexpected characteristics, which sometimes cannot be fully assessed experimentally. Theoretical approaches have been intensively employed in this particular case to investigate the systems at the atomic level. Beyond the analysis of observed phenomena, the prediction of unknown attributes has been attempted.

In this book-chapter, we review and discuss the current status and challenges of DFT studies of MOFs. We focus in particular on aspects relevant to catalysis at the metal centers found in these structures.

2. DFT calculations on MOFs

One of the main issues of DFT calculations is that a compromise between the dependability of the results and computational costs must be found, especially for large and complex systems [15]. A primary issue is how to select a functional suitable for the investigated system [16]. Different functionals, based on different approximations, may give different (even opposite) results. It is thus important to choose the functionals wisely. No functional devised up to now has proven to be suitable for all systems. Therefore, the validation of any selected functional is crucial.

MOFs are crystalline materials with extended frameworks. This needs, in some instances, to be taken into account when, for catalytic applications of MOFs, DFT is used to investigate the reaction mechanism. The reaction enthalpy can be obtained approximately from the energy at 0 K by including a zero-point energy correction calculated from (possibly scaled) vibrational frequencies. To compute the entropy, harmonic or anharmonic low-frequency modes can be used. The temperature-dependent free and activation energies can be calculated by adding thermal enthalpic and entropic effects. The so-obtained free energies of reaction are used to calculate temperature-dependent equilibrium constants. Note that the kinetics of a reaction can also be investigated, e.g., by transition state theory (TST) [17]. However, this is beyond the scope of this chapter.

Since the properties of MOFs depend not only on their chemical composition but also their structures, atomistic structural characterization is essentials. Generally, the structure of a newly synthesized MOF is characterized by single-crystal X-ray diffraction (SC-XRD). If the quality of the single crystals is not sufficient, the structure can be inferred from powder diffraction patterns [18] which often remains incomplete because of the complicated molecular system, large unit cell, and low symmetry. Molecular modeling methods must thus complement these experiments [19].

The periodic structure of an MOF can be studied by DFT: properties such as cell parameters and elastic properties can be obtained from periodic calculations. However, such calculations consume large amounts of computational resources due to the typically large unit cell of MOFs. Simple functionals such as local-density approximation (LDA) or generalized gradient approximations (GGA) are, however, often sufficient in this case [20]. Studying catalytic effects requires more advanced "hybrid" functionals, e.g., BP86 [21], B3-LYP [22], or M06 series [23].

Molecular mechanics (MM) is a much cheaper computational method to study large systems. The reliability of these calculations depends on the quality of the force fields used to describe the interactions between the atoms in the investigated system. Some high-accuracy force fields for MOFs, such as BTW-FF [24], MOF-FF [25], Quick-FF [26], and others [27, 28], were parametrized from DFT calculations. They were successfully used to determine and predict the periodic structures [29, 30], as well as to examine phase transformations [31–33], guest diffusion in the pores and other mechanical properties such as bulk moduli [34], elastic constants [35] and mechanical stability [36]. Note that MM calculations cannot describe electronic structures. Bond breaking and forming cannot be accounted for in such calculations. Thus, it is not possible to study reaction mechanisms using conventional MM methods [37].

In summary, for structurally well-defined and small enough systems, DFT calculations are a cogent approach especially in the field of localized catalytic reactions. We aim here to outline the suitability of DFT techniques to study catalysis in MOFs materials through examples from the recent literature.

3. Catalytic MOFs

MOFs have well-defined crystal structures with high concentrations of metal centers organized at topical distances, and large pore volumes between these centers. Even though MOFs are less able to withstand high temperatures compared to some other catalytic porous materials, their undeniable qualities such as large internal surface areas and uniform pore and cavity sizes make them attractive for various applications [38]. The pore size/shape in the framework can be tuned for selectivity for a particular reaction. In contrast, the small pore windows of other nanoporous materials, such as zeolites, commonly limit the catalytic transformation of large molecules [39].

In principle, various types of active sites can be incorporated into MOFs. Coordinatively unsaturated metal sites (CUSs) and functional groups on the organic linkers (usually acid/base sites) are the main catalytic sites. In addition, even though the MOF framework itself does not contain active sites, the catalytic reactivity can be enhanced by (a) postsynthetic functionalized modification [40] and (b) encapsulation of catalytic species [41]. The pore volume of MOFs is able to accommodate organic molecules, inorganic nanoparticles, metal complexes, and enzymes to conduct catalytic reactions. Here, we focus only catalytic reactions taking place on one CUS.

It is because of these functionalities [39] that MOFs were originally proposed for applications in catalysis. From this point of view, one tends to focus on the design and synthesis of MOFs with large pore sizes. The purpose is to allow the reactant to diffuse easily into the framework reaching the catalytic centers. Suitable synthetic techniques make it possible to choose the linkers and the metal-containing nodes to construct MOF suitable for particular reaction. Such a well-designed MOF catalyst should be highly reactive, selective, and stable. In order to approach this goal, some understanding of the reaction mechanisms is essential. In many cases, the reactants and resulting products tend to be well localized, for a sufficiently long time, in the framework due to host-guest interaction. The catalytic site can thus often be separated from the framework and investigated independently, making it easier to follow the reaction mechanisms [42].

We focus in this brief overview, mostly through a series of examples, on such local analyses of the activity of single-site catalysts. Selected examples of work based on cluster, or truncated, models of reactions catalyzed by metal centers, metal substitutions, and deposited metal complexes are succinctly reviewed.

4. Catalytically reactive metal nodes

As mentioned above, MOFs are built up from metal ions or metal-containing clusters as the inorganic building node and organic ligands as linkers. In some MOFs, coordinatively unsaturated metal sites (CUSs) have a catalytic effect. The sites are spatially well separated and, ideally, structurally identical. Thus, to investigate the reaction mechanism with DFT methods, truncated structures can often be used to represent the entire MOFs.

4.1. Reactivity of coordinatively unsaturated metal sites

MOFs with CUSs provide identical active metal sites, which are spatially isolated from each another. The sites are structurally well-characterized, without coupling to their neighbors. In typical dense heterogeneous catalyst, the reaction takes place on the outer surface [43]. One way to enhance the catalytic effect is to increase the accessible surface area. In MOFs, reactions occur not only on the outer surface, easily accessible active sites inside the framework also contribute.

Maihom et al. investigated the epoxidations of ethylene over $Fe_3(BTC)_2$ (BTC = 1,3,5-benzenetricarboxylate) using N_2O as oxidant [44]. **A** truncated model $[Fe_2(H_2BTC)_4]$ (**Figure 1a**) was used. The reaction mechanism started with the N_2O decomposition over Fe to generate an active Fe-oxo moiety, requiring an activation energy of 23.7 kcal/mol. An ethylenoxy intermediate, and eventually, ethylene oxide, were obtained as final products rather than acetaldehyde (**Figure 1b**).

An investigation of the epoxidation of propylene using $Cu_3(BTC)_2$ and $Fe_3(BTC)_2$ was recently carried out by the same group [45]. As mentioned, truncated clusters, $[Cu_2(H_2BTC)_4]$ and $[Fe_2(H_2BTC)_4]$, are suitable for these calculations, carried out at the M06-L/6-31G(d,p) level. As expected, the calculations showed that $Fe_3(BTC)_2$ is more active than $Cu_3(BTC)_2$ due to a

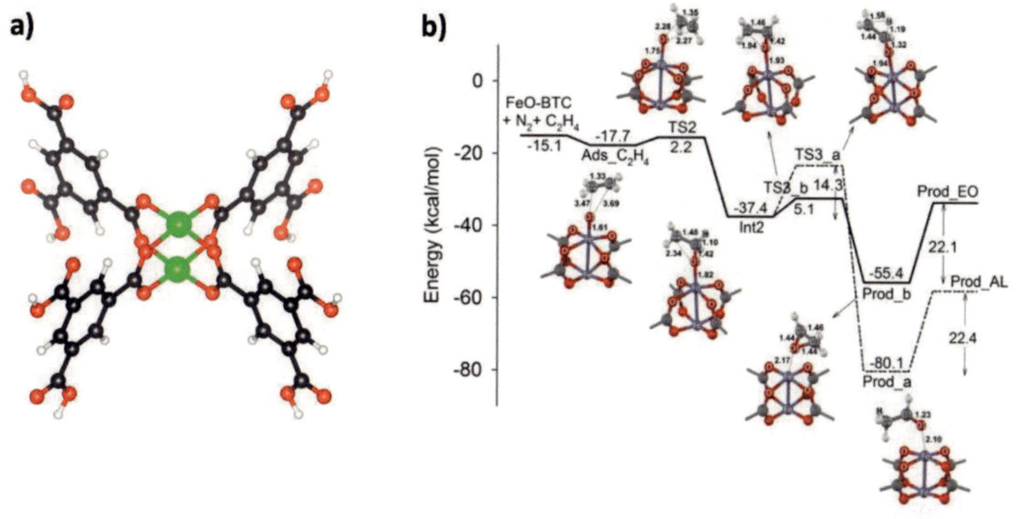

Figure 1. (a) Truncated model of $[Fe_2(H_2BTC)_4]$ and (b) energy profile of reactants, intermediates, and transition states involved in the formation of ethylene oxide (solid line) and the acetaldehyde (dash line) over the Fe paddle wheel. Adapted from Ref. [44] with permission of Wiley, copyright 2016.

larger charge transfer from the CUS to the oxidant O_2. It was found that the production of propylene oxide is favored over that of carbonylic products (propanal and acetone). Propanal and acetone were formed on the Fe-MOF cluster via the formation of a C—O bond. Then, the propyleneoxy intermediates and acetone are formed via a 1,2-hydride shift.

The M-MOF-74 series (M_2(DOBDC), where DOBDC = 2,5-dioxidoterephthalate and M = Mg, Ni, Co, Cu, and Zn) has been proposed as good catalysts for several reactions, owing to their reactive CUSs and large 1D channels, beneficial for the reactants' access to the active site. They are also thermally and chemically very stable [46–48]. Valvekens et al. [49] used MOF-74 with various metal ions, i.e., Mg (II), Ni (II), Co (II), Cu (II), and Zn (II) as Lewis acid catalysts to promote Knoevenagel condensations and Michael additions.

DFT calculations were performed on truncated models cut from the periodic geometries (**Figure 2**) optimized at the PBE-D2 level. The catalytic activities of M_2(DOBDC) systems with respect to the Knoevenagel condensation and Michael additions were examined. The calculations at the B3LYP level with a 6–31 g(d) basis showed that Ni-MOF-74 is the most active catalyst for both reactions. In addition, it was found that the phenolate groups coordinated with the CUSs substantially increase the catalytic performance. The phenolate oxygen proved to be a stronger base than the carboxylate oxygen, resulting in more acidic CUSs, enhancing the catalytic activity.

Llabrés i Xamena et al. [50] demonstrated that Cu(2-pymo)$_2$ and Co(PhIM)$_2$ (2-pymo and PhIM are 2-hydroxypyrimidinolat and phenylimidazolate, respectively) promote the aerobic oxidation reaction converting tetralin hydrocarbon to ketone and alcohol derivatives. The tetralin was first oxidized to hydroperoxides under oxygen condition and then decomposed to ketone and alcohol under the influence of the Lewis acid sites Cu(II) and Co(II). Ryan et al. [51] studied the mechanism of the hydroperoxide decomposition over three different complexes

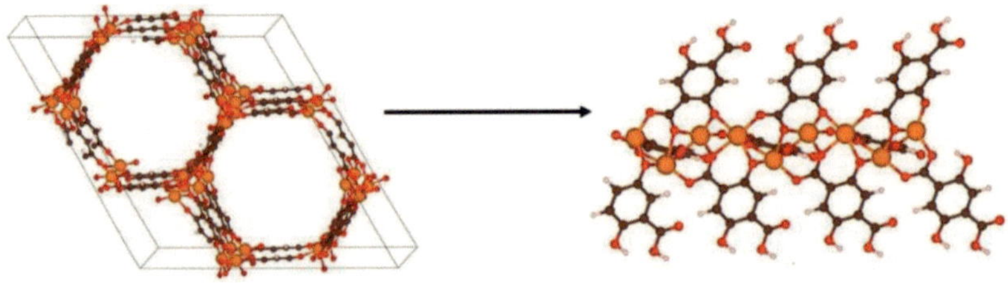

Figure 2. Periodic structure of M_2(DOBDC) optimized at the PBE-D2 level of theory (left), and the cluster model $M_9(DOBDC)_9$, truncated from the periodic structure (right). Adapted from Ref. [49] with permission of Elsevier, copyright 2014.

possessing copper or cobalt CUSs, i.e., Co(imidazolate)$_4$, Cu_2(2-hydroxypyrimidinolat)$_4$, and Cu_2(acetate)$_4$, as illustrated in **Figure 3**.

From DFT calculations (B3LYP/TZVP/6-31G(d,p)), the access of hydroperoxide to the CUSs of Cu_2(2-hydroxypyrimidinolat)$_4$ and of Co(imidazolate)$_4$ was found to be blocked by this bulky ligand. As a consequence, no decomposition of hydroperoxide was observed for the two systems. On the other hand, for Cu_2(acetate)$_4$, i.e., without the bulky ligands, the decomposition should occur. However, the calculations revealed that Cu_2(acetate)$_4$ was not active for the decomposition of tetralin. The energy barrier of the O—O bond cleavage over Cu_2(acetate)$_4$ (35.6–36.8 kcal/mol) was almost identical with that of the cleavage in the gas phase without catalyst (36.0–36.7 kcal/mol). The external surface was then considered as the active sites.

A complex consisting of Cu coordinated by three organic linkers and one water molecule was modeled representing the edge or outer surface of the framework (**Figure 4a**). The energy barrier of the hydroperoxide decomposition on the new model complex decreased to 25.3 kcal/mol, which was considerably lower than the value of the gas-phase reaction without MOFs as shown in **Figure 4b**.

Vanadium-based MIL-47(V), [VO(BDC)] (BDC = benzene-1,4-dicarboxylate), is an active catalyst for the liquid-phase cyclohexene oxidation-reaction using tert-butyl hydroperoxide (TBHP) as an oxidant [52]. MIL-47(V) is built up from linear V-(μ_2-O)-V chains interconnected

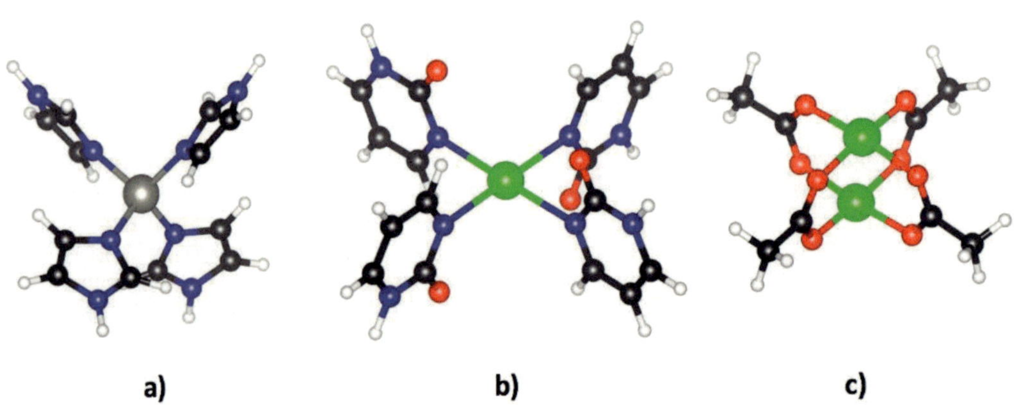

Figure 3. The metal cluster models taken from (a) Co(imidazolate)$_4$, (b) Cu_2(2-hydroxypyrimidinolat)$_4$, and (c) Cu_2(acetate)$_4$.

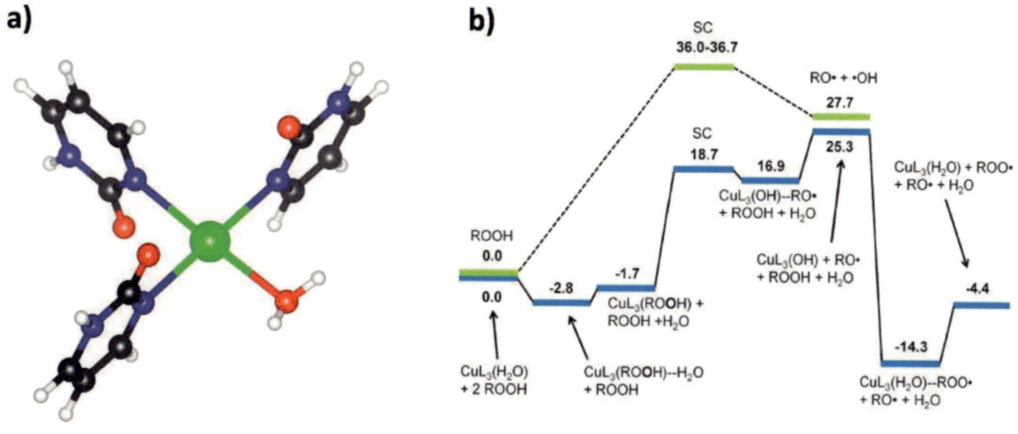

Figure 4. (a) Model of a cluster representing the exterior of a Cu_2(2-hydroxypyrimidinolat)$_3$ with one water molecule, (b) Gibbs free energy along the reaction coordinate for the proposed reaction cycle over the Cu_2(2-hydroxypyrimidinolat)$_3$ with one water molecule (blue) and oxygen—oxygen bond cleavage without catalyst (green). All energies are in kcal/mol. Figure adapted from Ref. [51] with permission of Elsevier, copyright 2012.

by terephthalate linkers to form a 3D framework possessing 1D rhombic channels. Each V(IV)-center in this chain coordinates to six oxygen atoms—four from carboxylate and two from μ_2O. Typically, MIL-47(V) does not provide CUSs. However, CUSs can be created from structural defects obtained by partially breaking the V—O_{COO} bonds and from the hydrolysis of the V—O_{COO} bonds.

The main products of the cyclohexene oxidation reaction are cyclohexene oxide, cyclohexane-1,2-diol, tert-butyl-2-cyclohexenyl-1-peroxide, and 2-cyclohexen-1-one with a conversion of about 80%. It was found that the catalytic performance of MIL-47(V) was comparable with that of typical homogeneous catalyst, $VO(acac)_2$. Note that no leaching of V(IV) was observed when the oxidant was dissolved in the solution. The structures of the MIL-47(V) were maintained intact until the end of reaction. Leus and Vandichel et al. [52] proposed a reaction pathway from DFT calculations on finite clusters of MIL-47(V) (**Figure 5**).

The results were in good agreement with EPR and NMR measurement that are interpreted in terms of the existence of (at least) two different catalytic pathways, described as the "direct" and "radical" pathways [52]. The first step of both pathways was the formation of a vanadium hydroperoxide species from TBHP. For the direct pathway, cyclohexene was directly converted to cyclohexene oxide, while the oxidation state of vanadium (V(IV)) did not change during the reaction. In the case of the radical pathway, V(IV) was oxidized to V(V), which further reacted with the oxidant TBHP to give an active species of the cyclohexene epoxidation. Finally, the catalyst was regenerated by interacting with TBHP again. At the B3LYP-D3 and 311 + g(3df,2p) level of theory, the reaction free energy for epoxidation reaction was 37 kJ/mol for the more favorable radical route.

Moreover, Matthias et al. [53] studied the cyclohexene oxidation using MIL-47(V)-functionalized linkers (with the functional groups —OH, —F, —Cl, —Br, —CH_3, and —NH_2) by means of experimental and theoretical works. Different catalytic conversions were observed depending on the functionalization of the linkers. DFT calculations on truncated clusters of MIL-47(V) and its derivatives were performed as in the abovementioned example for the radical

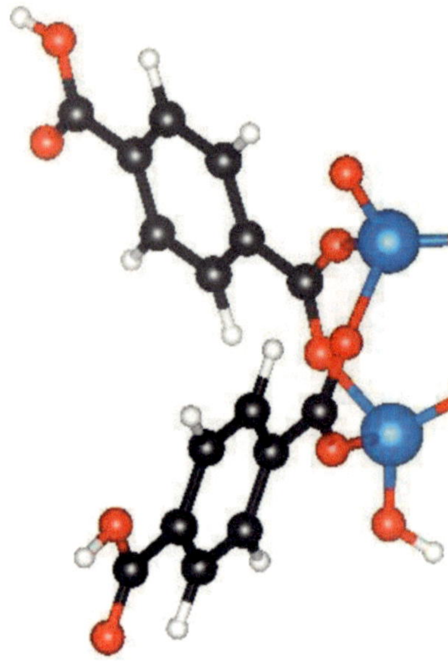

Figure 5. Simulated model of V-MIL-47.

pathway. The results showed that, compared to the parent MIL-47(V), the catalytic efficiency significantly increased due to the functional groups. Among these, the —OH-functionalized structure showed the largest improvement (lowest activation energy), which was in good agreement with the experimental results. This is because of a strong hydrogen bond between the —OH and the alkylperoxo group possibly stabilizing the transition state.

4.2. Reactivity of substituted metal centers as single-site catalysts

Partial substitution of metal ions while maintaining the structural integrity and the porosity is a promising strategy enabling some otherwise inactive MOFs to show a catalytic activity [54]. The metal substitution usually involves the cleavage and regeneration of coordination bonds between metal ions and organic ligands. The atomic-level understanding of this process is difficult, making room for molecular modeling. Here, we show examples of theoretical studies on the catalytic activity of metal-substituted MOFs.

Xiao et al. [55] reported the oxidation reaction of ethane to ethanol over magnesium-diluted $Fe_{0.1}Mg_{1.9}$(DOBDC), MOF-74, in which 5% of the redox-inactive Mg(II) were substituted by the active Fe(II). This framework provides wide hexagonal channels beneficial for an easy access of the reactants to the site-isolated open Fe-CUSs. Nitrous oxide (N_2O) was used as the oxidant, generating highly reactive Fe(IV)-oxo intermediates, which could further activate strong C—H bonds of alkanes, yielding ethanol and acetaldehyde. Owing to the short lifetime of Fe(IV)-oxo, it could neither be isolated nor characterized by conventional experiments.

Shortly afterward, Verma el al. [56] applied DFT calculations to confirm whether the Fe(IV)-oxo species are formed and act as active centers for ethane hydroxylation to ethanol. They modeled the Fe sites in magnesium-diluted Fe_2(DOBDC) by replacing 1 Mg atom with Fe atom in cluster models comprising 88 atoms of Mg-MOF-74. The oxidation of ethane to ethanol over this magnesium-diluted MOFs using N_2O as oxidant was proposed to proceed via a two-step consecutive reaction (**Figure 6**): the formation of the Fe(IV)-oxo unit via N_2O decomposition and the hydroxylation of ethane to ethanol over the Fe(IV)-oxo complex. It is found that the first step is the rate-determining step with an activation energy of 82 kJ/mol. Compared to the uncatalyzed reaction, where nitrous oxide directly oxidizes the ethane to ethanol, it was found that the reaction requires an activation barrier of 280 kJ/mol. This suggests that the Fe(IV)-oxo species is indeed an active center for the oxidation of ethane to ethanol.

In the meantime, Hirrao et al. [57] also carried out combined quantum and molecular mechanics (QM/MM) calculations for the hydroxylation of ethane, in analogy to Mg-MOF-74. This methodology treats the active site by QM and the rest of the system is treated by MM. The accuracy of QM/MM methods certainly depends on the choice of the QM theory level and the set of MM parameters, also on the treatment of electrostatics at the interface between the QM and MM regions. One of three Mg(II) ions was substituted by Fe(II) in the QM region, as in the model suggested by Verma et al. [56]. In the ONIOM [58] scheme, the MM region was treated at the B3LYP/[SDD(Fe),6-31G(d)(others)] level while the UFF interaction model [59] was used for the MM region. The 6–311 + G(df,p) basis set was used for single-point reference calculations. This work also suggested that the spin state (S = 2) of the Fe(IV)-oxo species does not change during the reaction.

Interestingly, Liao et al. [60] then improved based on Verma et al. model [56] the understanding on the activity of Fe open sites for the oxidation of ethane reaction. They studied MOF-74 with different linkers, in particular, $-CH_3$, NH_2, $-COOH$, $-CN$, $-OH$, $-OCH_3$, $-N(CH_3)_2$,

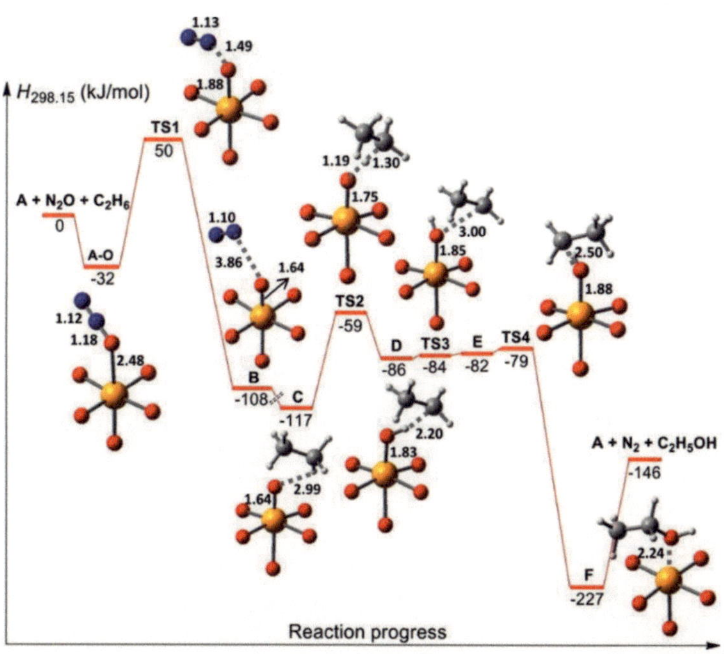

Figure 6. Enthalpy profile, $\Delta H_{298.15}$ (in kJ/mol), for the intermediates and transition states of the catalytic cycle. The key bond distances are given in Å (color code: orange = Fe, red = O, blue = N, gray = C, and white = H). Reproduced from Ref. [56] with permission of the American Chemical Society, copyright 2015.

and 4-pyridyl functionalized groups on different linkers such as using methanol, pyridine, formate, benzoate anion, and the imidazole anion. The authors concluded that the population of the d-orbital was significantly influenced by the coordination ligand field. The 3d orbital energy of Fe correlated with the electron-donating strength of the functional groups on the linkers. The results reveal that linkers with $-NH_2$ groups reduce the enthalpic barrier for the most endothermic step in the ethane oxidation pathway. This study illustrates the use of simple models to understand complicated and computationally intensive systems. The activity of CUSs thus might be optimized by selecting suitable ligand environments, which might be useful for upgrading certain hydrocarbon process.

The design of Fe(IV)-oxo complexes in MOFs was further reported by Impeng et al. [61]. In this work, we demonstrated the possibility of designing Fe-oxo complexes in MOFs for the activation of alkane C—H bond by incorporating an Fe ion into a Zn-based cluster derived from $Zn_4O(BDC)_6$, known as MOF-5 [62], and generating the Fe-O unit through N_2O dissociation on an Fe-substituted Zn-based cluster (Fe-Zn_3O(pyrazole)$_6$). The calculations with B3LYP-D3 showed that both steps are feasible and that the catalytic activity of Fe-Zn_3O(pyrazole)$_6$ for N_2O decomposition is on a par with the Fe sites in magnesium-diluted Fe_2(DOBDC). Concerning ethane C—H bond activation, in addition to σ and π pathways on triplet and quintet surfaces, an alternative unusual pathway, called δ, is also observed on the triplet surface. The σ pathway on the quintet surface (TS^5_σ) has the lowest activation energy owing to less steric hindrance and favorable d-d interactions on the Fe active site at the transition state, as illustrated in **Figure 7**.

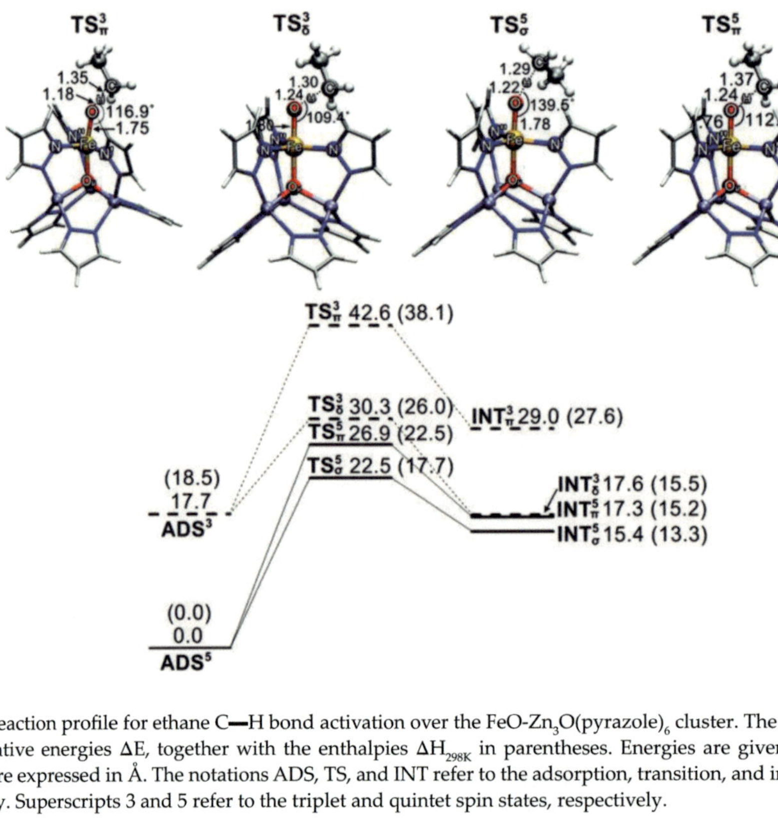

Figure 7. Reaction profile for ethane C—H bond activation over the FeO-Zn_3O(pyrazole)$_6$ cluster. The inserted numbers are the relative energies ΔE, together with the enthalpies ΔH_{298K} in parentheses. Energies are given in kcal/mol and distances are expressed in Å. The notations ADS, TS, and INT refer to the adsorption, transition, and intermediate steps, respectively. Superscripts 3 and 5 refer to the triplet and quintet spin states, respectively.

5. Reactivity of metal nodes deposited as single-site catalysts

In this section, we discuss systems possessing metal-complexes/ions anchored on the inorganic building nodes of the framework, which can be achieved by postsynthetic techniques such as metalation [63], cation exchange [64], or atomic layer deposition [65]. Zr-based MOFs consist of Zr_6O_8 as a primary node linked with carboxylate linkers. The Zr_6O_8 nodes connected by 12 linkers result in the UiO-66 series, while the partial connection with 8 and 6 carboxylates produces NU-1000 and MOF-808, respectively. The catalytic reactivity can be enhanced through such modifications. This was often employed for Zr-based MOFs, which are remarkable because of their appealing chemical and thermal stability [40]. Well-known Zr-based MOFs such as MOF-808 $[Zr_6(\mu_3\text{-}O)_4(\mu_3\text{-}OH)_4(OH)_6(H_2O)_6(BTC)_2]$, NU-1000 $[Zr_6(\mu_3\text{-}O)_4(\mu_3\text{-}OH)_4(OH)_4(H_2O)_4(TBAPy)_2]$ (TBAPy = tetratopic 1,3,6,8-tetrakis(p-benzoate)pyrene) and UiO-66, known as NU-1000 and $[Zr_6(\mu_3\text{-}O)_4(\mu_3\text{-}OH)_4(BDC)_6]$, known as UiO-66 have been studied intensively [66, 67].

Ortuño et al. [68] reported on a computational screening of the first row of divalent transition metals (i.e., Fe(II), Co(II), Ni(II), Cu(II), and Zn(II)) supported on NU-1000 (**Figure 8**) for acceptorless alcohol dehydrogenation. The author proposed the reaction taking place via a three-step reaction mechanism, composed of (i) a proton transfer, (ii) β-hydride elimination, and (iii) H—H bond formation. The Fe(II), Co(II), and Ni(II) complexes, consistent with weak-field oxide ligands, had high-spin ground electronic states as quintet, quartet, and triplet, respectively. The Cu(II) and Zn(II) species were predicted to have doublet and singlet ground states, respectively. It was found that the Co(II) and Ni(II) supported NU-1000 were the two most promising catalysts for the acceptorless alcohol dehydrogenation with an activation free energy of 28.5 and 26.5 kcal/mol, respectively.

Later, the same group also studied Ni(II) and Co(II) deposited on Zr-NU-1000 as a catalyst for ethylene dimerization, which converts ethylene to 1-butene and 2-butene [69]. For the structure optimization, the DFT level of theory (M06-L) and def2-TZVPP basis set were employed, whereas complete active space self-consistent field (CASSCF) and second-order perturbation theory (CASPT2) were used for the electronic structure characterization of the reactive species. The NU-1000 models were used as the same as in the previous example (**Figure 8**). The grafted Ni and Co were terminated with active —OH and —OH_2 groups.

The Cossee-Arlman mechanism was found as the energetically preferable pathway. Ethylene insertion into the existing metal-ethyl bond was the rate-determining step. Concerning the spin state of the two catalysts, the Co(II) species can have doublet or quartet states and the Ni(II) can have singlet or triplet states. Based on these calculations, Ni(II)-modified NU-1000 (with an activation energy of 15.6 kcal/mol) presents a greater reactivity than the Co(II)-modified system (with an activation energy of 24.1 kcal/mol) for the ethylene insertion into the metal-carbon bond. The energetics of ethylene dimerization for the Ni(II) and Co(II) complexes are shown in **Figure 9**. From the most relevant CASSCF molecular orbitals (MOs), the *3d* orbitals of the low-spin Ni(II) complex hybridized with the *2p* orbitals of the active carbon atom because of the empty fifth orbital. In contrast to the high-spin Co(II) complex, the degree of hybridization in the Ni-case is less able to stabilize the TS structure due to the half-filled *3d* orbital, leading to a lower catalytic activity.

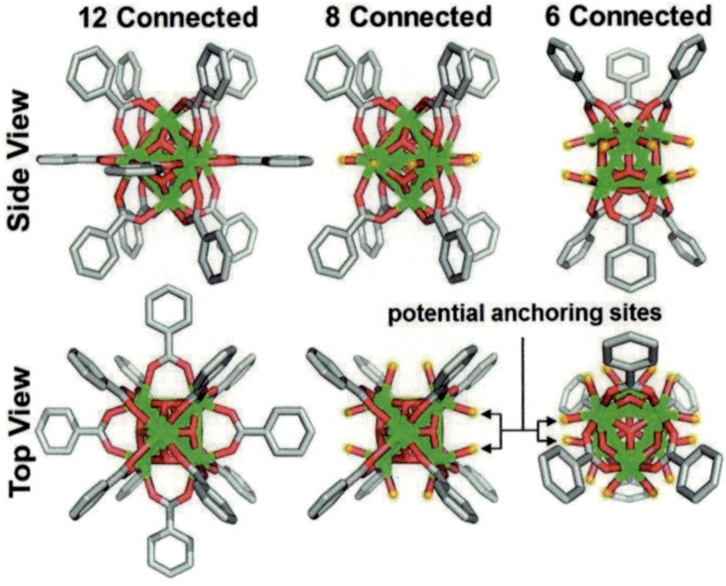

Figure 8. The "top" and "side" views of a 12-, 8-, and 6-connected Zr_6 oxide node. The arrows indicate the unsaturated 8- and 6-connected nodes, which have potential anchoring points. Reproduced from Ref. [66] with permission of Wiley, copyright 2016.

The oxidative dehydrogenation mechanism of propane on an active Co(II)-oxygen cluster anchored on the inorganic node of Zr-NU-1000 [70] was also explored by DFT with the M06-L functional and def2-TZVPP. A highly selective propane dehydrogenation at low temperature (230°C) has been found in experiments. Li et al. [70] found a computed propene formation

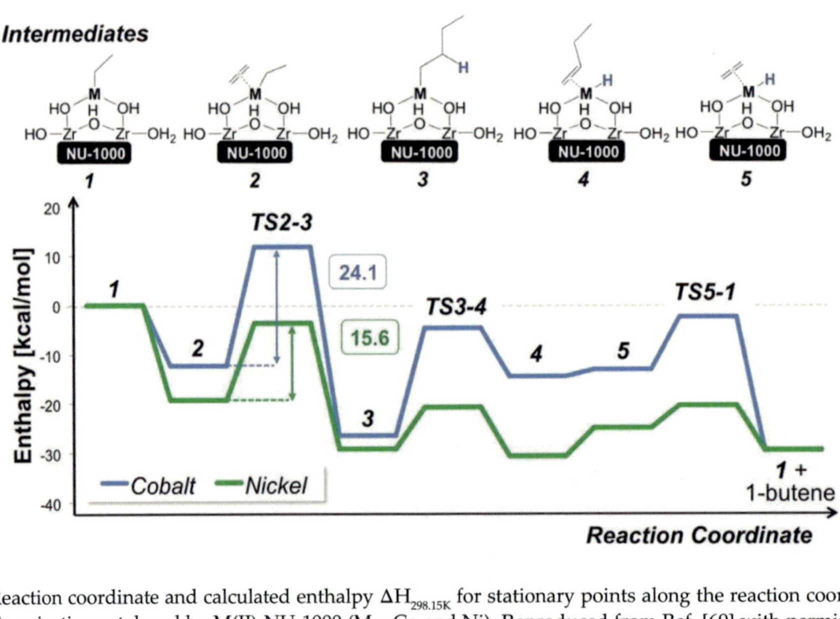

Figure 9. Reaction coordinate and calculated enthalpy $\Delta H_{298.15K}$ for stationary points along the reaction coordinates for ethylene dimerization catalyzed by M(II)-NU-1000 (M = Co and Ni). Reproduced from Ref. [69] with permission of the American Chemical Society, copyright 2016.

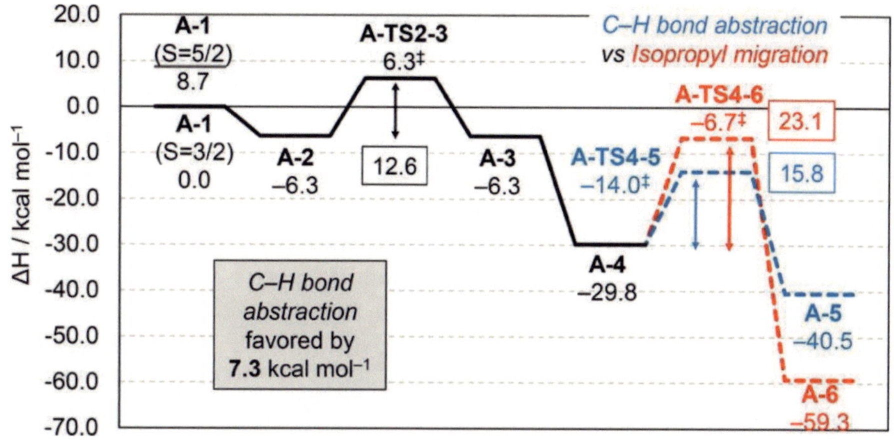

Figure 10. Computed enthalpies, ΔH_{503K} (kcal/mol), for propane oxidative dehydrogenation. Reproduced from Ref. [70] with permission of the American Chemical Society, copyright 2017.

in line with experimental results. DFT cluster calculations allowed to study the reaction and showed the influence on the reaction of adding Co(II) cations as a secondary metal. The computed enthalpies for the equilibrium structures are shown in **Figure 10**.

In more detail, the reactive Co(III)—O˙ moiety was generated by the catalyst regeneration reaction with O_2 having a strong interaction with the propane molecules, promotes the propane dehydrogenation, and forms isopropyl at the Co(III) active site. The abstraction of the terminal C-H bond by the Co(III)—O˙ intermediate is the rate-determining step, yielding

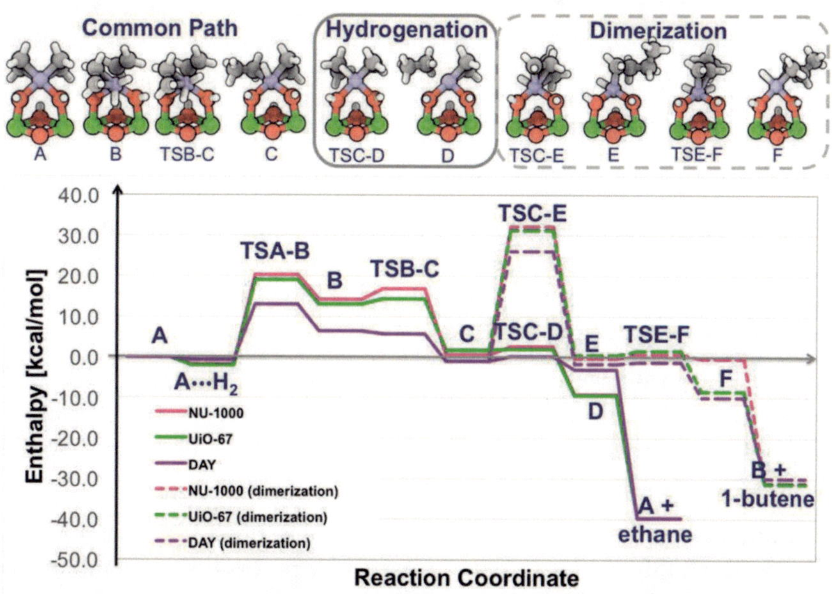

Figure 11. Calculated enthalpy, $\Delta H_{298.15K}$ (kcal/mol), for ethylene hydrogenation to ethane and dimerization to 1-butene catalyzed by supported metal complexes initially present as $Rh(C_2H_4)_2$ on NU-1000, UiO-67, or DAY zeolites. Reproduced from Ref. [71] with permission of the American Chemical Society, copyright 2017.

Co(II)—OH and isopropyl radicals, which are converted into propene. The energy required for the H abstraction producing propene (15.8 kcal/mol) was significantly lower than that for the isopropyl migration (23.1 kcal/mol), the propene formation path is thus preferable.

Gates et al. [71] showed that Rh(I) and Ir(I) diethylene complexes can be anchored on the Zr-NU-1000, Hf-NU-1000, and Zr-UiO-67 to catalyze ethylene hydrogenation and dimerization. They also used DAY zeolite as supporting material to compare the catalytic performance. DFT calculations with M06-L and def2-TZVPP were employed. The optimized geometries were in good agreement with the data analyzed from IR spectra and X-ray absorption fine structure (EXAFS) results. The ethylene insertion to the ethyl—Rh bond (rate-determining step) significantly depended on the nature of the supporting MOF. The catalytic activity was compared between Zr- and Hf-nodes and DAY zeolites. The calculated enthalpies for the ethylene hydrogenation and dimerization were shown in **Figure 11**. The calculations reveal a lower energy barrier of the DAY zeolite as supporting material than Zr-NU-1000, Hf-NU-1000, and Zr-UiO-67 for the hydrogenation and dimerization. This is opposite to the experiments which show a better catalytic activity of the DAY zeolite than the MOFs materials. The authors proposed that this might be because of the spillover effect rather than a direct electronic effect on the Rh(I) complex.

6. Outlook

We focus in this brief overview on theoretical studies of the catalytic activity of single-site catalytic MOFs. DFT-based approaches can provide a detailed, if only local, description of the chemistry occurring in various environments encountered in such MOFs. This not only helps in the interpretation of experiment data; the abilities of DFT calculations are a crucial characterization method toward a deeper molecular perspective. This technique is now powerful and reliable enough to predict and guide the synthesis of materials.

There are, however, several problems which need to be overcome:

I. To find a link between a predicted structure and its syntheses still a challenge.

II. Sometime nonlocal phenomena such as a stimulus-responsive behavior of the catalytic site itself must be taken into account. Many works[ref] attempt to simulate larger models in order to describe such systems. This seems to require improved methodologies to limit the computational requirements.

III. The functionals: Since there is no "universal functional" DFT must be always used with a lot of caution. A lot of experience has been accumulated on the suitability of various approximations to investigate MOFs. Yet new, and costly, validations are still required whenever a "new" system is to be studied.

IV. Last but not least, in some cases, a local view as described in this chapter may not be sufficient to catch the main features of a particular process. The overall outcome of the process under study may depend on many factors, in the worst case, none of which dominant. Various "local views" that may have been obtained could be combined as building blocks (e.g., by understanding the molecular motions between "sites"). This remains largely a task for theoretical improvements in the future.

Acknowledgements

The writing and publication of this book chapter was financially supported by Vidyasirimedhi Institute of Science and Technology and the Thailand Research Fund (TRF) (RSA6080068).

Conflicts of interest

There are no conflicts to declare.

Author details

Siwarut Siwaipram[1], Sarawoot Impeng[2], Philippe A. Bopp[1,3] and Sareeya Bureekaew[1*]

*Address all correspondence to: sareeya.b@vistec.ac.th

1 Department of Energy Science and Engineering (ESE), Vidyasirimedhi Institute of Science and Technology (VISTEC), Rayong, Thailand

2 National Nanotechnology Center (NANOTEC), National Science and Technology Development Agency, Pathumthani, Thailand

3 Department of Molecular Science and Engineering (MSE), Vidyasirimedhi Institute of Science and Technology (VISTEC), Rayong, Thailand

References

[1] Anastas PT, Kirchhoff MM. Origins, current status, and future challenges of green chemistry. Accounts of Chemical Research. 2002;**35**(9):686-694. DOI: 10.1021/ar010065m

[2] Law AL. Simulation Modeling and Analysis. Vol. 2. New York: McGraw. McGraw-Hill New York; 2007. Retrieved from https://www.researchgate.net/

[3] Whitehead TL. Molecular modeling: Basic principles and applications, 2nd Edition (Hans-Dieter Höltje, Wolfgang Sippl, Didier Rognan, and Gerd Folkers). Journal of Chemical Education. 2006;**83**(6):851. DOI: 10.1021/ed083p851

[4] Becke AD. Density-functional thermochemistry. III. The role of exact exchange. The Journal of Chemical Physics. 1993;**98**(7):5648-5652. DOI: 10.1063/1.464913

[5] Pople JA, Gill PMW, Johnson BG. Kohn-Sham density-functional theory within a finite basis set. Chemical Physics Letters. 1992;**199**(6):557-560. DOI: 10.1016/0009-2614(92)85009-y

[6] Furukawa H, Cordova KE, O'Keeffe M, Yaghi OM. The chemistry and applications of metal-organic frameworks. Science. 2013;**341**(6149):123044-0-123044-12. DOI: 10.1126/science.1230444

[7] Férey G. Hybrid porous solids: Past, present, future. Chemical Society Reviews. 2008;**37**(1):191-214. DOI: 10.1039/b618320b

[8] Kitagawa S, Matsuda R. Chemistry of coordination space of porous coordination polymers. Coordination Chemistry Reviews. 2007;**251**(21-24):2490-2509. DOI: 10.1016/j.ccr.2007.07.009

[9] Murray LJ, Dincă M, Long JR. Hydrogen storage in metal-organic frameworks. Chemical Society Reviews. 2009;**38**(5):1294-1314. DOI: 10.1039/b802256a

[10] Li J-R, Kuppler RJ, Zhou H-C. Selective gas adsorption and separation in metal-organic frameworks. Chemical Society Reviews. 2009;**38**(5):1477-1504. DOI: 10.1039/b802426j

[11] Rogge SMJ, Bavykina A, Hajek J, Garcia H, Olivos-Suarez AI, Sepúlveda-Escribano A, et al. Metal-organic and covalent organic frameworks as single-site catalysts. Chemical Society Reviews. 2017;**46**(11):3134-3184. DOI: 10.1039/c7cs00033b

[12] Horcajada P, Serre C, Vallet-Regí M, Sebban M, Taulelle F, Férey G. Metal-organic frameworks as efficient materials for drug delivery. Angewandte Chemie. 2006;**118**(36):6120-6124. DOI: 10.1002/anie.200601878

[13] Horcajada P, Gref R, Baati T, Allan PK, Maurin G, Couvreur P, et al. Metal-organic frameworks in biomedicine. Chemical Reviews. 2012;**112**(2):1232-1268. DOI: 10.1021/cr200256v

[14] Hu Z, Deibert BJ, Li J. Luminescent metal-organic frameworks for chemical sensing and explosive detection. Chemical Society Reviews. 2014;**43**(16):5815-5840. DOI: 10.1039/C4CS00010B

[15] Sholl D, Steckel JA. Density Functional Theory: A Practical Introduction. John Wiley & Sons; 2009. pp. 1-33. DOI: 10.1002/anie.200905551

[16] Mattsson AE, Schultz PA, Desjarlais MP, Mattsson TR, Leung K. Designing meaningful density functional theory calculations in materials science—A primer. Modelling and Simulation in Materials Science and Engineering. 2004;**13**(1):R1. DOI: 10.1088/0965-0393/13/1/R01

[17] Steinfeld JI, Francisco JS, Hase WL. Chemical Kinetics and Dynamics. Vol. 3. New Jersey: Prentice Hall Englewood Cliffs; 1989. Retrieved from: http://www.academia.edu/

[18] Price SL. From crystal structure prediction to polymorph prediction: Interpreting the crystal energy landscape. Physical Chemistry Chemical Physics. 2008;**10**(15):1996-2009. DOI: 10.1039/b719351c

[19] Mellot-Draznieks C. Role of computer simulations in structure prediction and structure determination: From molecular compounds to hybrid frameworks. Journal of Materials Chemistry. 2007;**17**(41):4348-4358. DOI: 10.1039/B702516P

[20] Kohn W, Becke AD, Parr RG. Density functional theory of electronic structure. The Journal of Physical Chemistry. 1996;**100**(31):12974-12980. DOI: 10.1021/jp960669l

[21] Becke AD. Density-functional exchange-energy approximation with correct asymptotic behavior. Physical Review A. 1988;**38**(6):3098-3100. DOI: 10.1103/PhysRevA.38.3098

[22] Lee C, Yang W, Parr RG. Development of the Colle-Salvetti correlation-energy formula into a functional of the electron density. Physical Review B. 1988;**37**(2):785-789. DOI: 10.1103/PhysRevB.37.785

[23] Zhao Y, Truhlar DG. The M06 suite of density functionals for main group thermochemistry, thermochemical kinetics, noncovalent interactions, excited states, and transition elements: Two new functionals and systematic testing of four M06-class functionals and 12 other functionals. Theoretical Chemistry Accounts. 2008;**120**(1):215-241. DOI: 10.1007/s00214-007-0310-x

[24] Bristow JK, Tiana D, Walsh A. Transferable force field for metal-organic frameworks from first-principles: BTW-FF. Journal of Chemical Theory and Computation. 2014;**10**(10):4644-4652. DOI: 10.1021/ct500515h

[25] Bureekaew S, Amirjalayer S, Tafipolsky M, Spickermann C, Roy TK, Schmid R. MOF-FF-A flexible first-principles derived force field for metal-organic frameworks. Physica Status Solidi. 2013;**250**(6):1128-1141. DOI: 10.1002/pssb.201248460

[26] Vanduyfhuys L, Vandenbrande S, Verstraelen T, Schmid R, Waroquier M, Van Speybroeck V. QuickFF: A program for a quick and easy derivation of force fields for metal-organic frameworks from ab initio input. Journal of Computational Chemistry. 2015;**36**(13):1015-1027. DOI: 10.1002/jcc.23877

[27] Coupry DE, Addicoat MA, Heine T. Extension of the universal force field for metal-organic frameworks. Journal of Chemical Theory and Computation. 2016;**12**(10):5215-5225. DOI: 10.1021/acs.jctc.6b00664

[28] Fang H, Demir H, Kamakoti P, Sholl DS. Recent developments in first-principles force fields for molecules in nanoporous materials. Journal of Materials Chemistry A. 2014;**2**(2):274-291. DOI: 10.1039/C3TA13073H

[29] Impeng S, Cedeno R, Dürholt JP, Schmid R, Bureekaew S. Computational structure prediction of (4, 4)-connected copper paddle-wheel-based MOFs: Influence of ligand functionalization on the topological preference. Crystal Growth & Design. 2018;**18**(5):2699-2706. DOI: 10.1021/acs.cgd.8b00238

[30] Bureekaew S, Balwani V, Amirjalayer S, Schmid R. Isoreticular isomerism in 4, 4-connected paddle-wheel metal-organic frameworks: Structural prediction by the reverse topological approach. CrystEngComm. 2015;**17**(2):344-352. DOI: 10.1039/C4CE01574F

[31] Alaghemandi M, Schmid R. Model study of thermoresponsive behavior of metal-organic frameworks modulated by linker functionalization. Journal of Physical Chemistry C. 2016;**120**(12):6835-6841. DOI: 10.1021/acs.jpcc.5b12331

[32] Wieme J, Vanduyfhuys L, Rogge SMJ, Waroquier M, Van Speybroeck V. Exploring the flexibility of MIL-47 (V)-type materials using force field molecular dynamics simulations. Journal of Physical Chemistry C. 2016;**120**(27):14934-14947. DOI: 10.1021/acs.jpcc.6b04422

[33] Evans JD, Bocquet L, Coudert FX. Origins of negative gas adsorption. Chem. 2016;**1**(6): 873-886. DOI: 10.1016/j.chempr.2016.11.004

[34] Boyd PG, Moosavi SM, Witman M, Smit B. Force-field prediction of materials properties in metal-organic frameworks. Journal of Physical Chemistry Letters. 2017;**8**(2):357-363. DOI: 10.1021/acs.jpclett.6b02532

[35] Ortiz AU, Boutin A, Fuchs AH, Coudert F-X. Metal-organic frameworks with wine-rack motif: What determines their flexibility and elastic properties? The Journal of Chemical Physics. 2013;**138**(17):174703. DOI: 10.1063/1.4802770

[36] Rogge SMJ, Wieme J, Vanduyfhuys L, Vandenbrande S, Maurin G, Verstraelen T, et al. Thermodynamic insight in the high-pressure behavior of UiO-66: Effect of linker defects and linker expansion. Chemistry of Materials. 2016;**28**(16):5721-5732. DOI: 10.1021/acs.chemmater.6b01956

[37] Van Duin ACT, Dasgupta S, Lorant F, Goddard WA. ReaxFF: A reactive force field for hydrocarbons. The Journal of Physical Chemistry. A. 2001;**105**(41):9396-9409. DOI: 10.1021/jp004368u

[38] Howarth AJ, Liu Y, Li P, Li Z, Wang TC, Hupp JT, et al. Chemical, thermal and mechanical stabilities of metal-organic frameworks. Nature Reviews Materials. 2016;**1**(3):15018. DOI: 10.1038/natrevmats.2015.18

[39] Kitagawa S, Kitaura R, Noro S. Functional porous coordination polymers. Angewandte Chemie, International Edition. 2004;**43**(18):2334-2375. DOI: 10.1002/anie.200300610

[40] Cohen SM. Postsynthetic methods for the functionalization of metal-organic frameworks. Chemical Reviews. 2011;**112**(2):970-1000. DOI: 10.1021/cr200179u

[41] Cui Y, Li B, He H, Zhou W, Chen B, Qian G. Metal-organic frameworks as platforms for functional materials. Accounts of Chemical Research. 2016;**49**(3):483-493. DOI: 10.1021/acs.accounts.5b00530

[42] Liu J, Chen L, Cui H, Zhang J, Zhang L, Su CY. Applications of metal-organic frameworks in heterogeneous supramolecular catalysis. Chemical Society Reviews. 2014;**43**(16): 6011-6061. DOI: 10.1039/C4CS00094C

[43] Thomas JM, Raja R, Lewis DW. Single-site heterogeneous catalysts. Angewandte Chemie, International Edition. 2005;**44**(40):6456-6482. DOI: 10.1002/anie.200462473

[44] Maihom T, Choomwattana S, Wannakao S, Probst M, Limtrakul J. Ethylene epoxidation with nitrous oxide over Fe-BTC metal-organic frameworks: A DFT study. Chemphyschem. 2016;**17**(21):3416-3422. DOI: 10.1002/cphc.201600836

[45] Maihom T, Sawangphruk M, Probst M, Limtrakul J. A computational study of the catalytic aerobic epoxidation of propylene over the coordinatively unsaturated metal-organic framework Fe3(btc)2: Formation of propylene oxide and competing reactions. Physical Chemistry Chemical Physics. 2018;**20**(9):6726-6734. DOI: 10.1039/c7cp07550b

[46] Xiao DJ, Oktawiec J, Milner PJ, Long JR. Pore Environment Effects on Catalytic Cyclohexane Oxidation in Expanded Fe2 (dobdc) Analogues. Journal of the American Chemical Society. 2016;**138**(43):14371-14379. DOI: 10.1021/jacs.6b08417

[47] Zuluaga S, Fuentes-Fernandez EMA, Tan K, Xu F, Li J, Chabal YJ, et al. Understanding and controlling water stability of MOF-74. Journal of Materials Chemistry A. 2016;**4**(14): 5176-5183. DOI: 10.1039/C5TA10416E

[48] Calleja G, Sanz R, Orcajo G, Briones D, Leo P, Martínez F. Copper-based MOF-74 material as effective acid catalyst in Friedel-Crafts acylation of anisole. Catalysis Today. 2014;**227**:130-137

[49] Valvekens P, Vandichel M, Waroquier M, Van Speybroeck V, De Vos D. Metal-dioxidoterephthalate MOFs of the MOF-74 type: Microporous basic catalysts with well-defined active sites. Journal of Catalysis. 2014;**317**:1-10. DOI: 10.1016/j.jcat.2014.06.006

[50] i Xamena FXL, Casanova O, Tailleur RG, Garcia H, Corma A. Metal organic frameworks (MOFs) as catalysts: A combination of Cu2+ and Co2+ MOFs as an efficient catalyst for tetralin oxidation. Journal of Catalysis. 2008;**255**(2):220-227. DOI: 10.1016/j.jcat.2008.02.011

[51] Ryan P, Konstantinov I, Snurr RQ, Broadbelt LJ. DFT investigation of hydroperoxide decomposition over copper and cobalt sites within metal-organic frameworks. Journal of Catalysis. 2012;**286**:95-102. DOI: 10.1016/j.jcat.2011.10.019

[52] Leus K, Vandichel M, Liu YY, Muylaert I, Musschoot J, Pyl S, et al. The coordinatively saturated vanadium MIL-47 as a low leaching heterogeneous catalyst in the oxidation of cyclohexene. Journal of Catalysis. 2012;**285**(1):196-207. DOI: 10.1016/j.jcat.2011.09.014

[53] Vandichel M, Biswas S, Leus K, Paier J, Sauer J, Verstraelen T, et al. Catalytic performance of vanadium MIL-47 and linker-substituted variants in the oxidation of cyclohexene: A combined theoretical and experimental approach. ChemPlusChem. 2014;**79**(8): 1183-1197. DOI: 10.1002/cplu.201402007

[54] Han Y, Li J-R, Xie Y, Guo G. Substitution reactions in metal-organic frameworks and metal-organic polyhedra. Chemical Society Reviews. 2014;**43**(16):5952-5981. DOI: 10.1039/C4CS00033A

[55] Xiao DJ, Bloch ED, Mason JA, Queen WL, Hudson MR, Planas N, et al. Oxidation of ethane to ethanol by N2O in a metal-organic framework with coordinatively unsaturated iron(II) sites. Nature Chemistry. 2014;**6**(7):590-595. DOI: 10.1038/nchem.1956

[56] Verma P, Vogiatzis KD, Planas N, Borycz J, Xiao DJ, Long JR, et al. Mechanism of oxidation of ethane to ethanol at Iron(IV)-oxo sites in magnesium-diluted Fe_2(dobdc). Journal of the American Chemical Society. 2015;**137**(17):5770-5781. DOI: 10.1021/jacs.5b00382

[57] Hirao H, Ng WKH, Moeljadi AMP, Bureekaew S. Multiscale model for a metal-organic framework: High-spin rebound mechanism in the reaction of the oxoiron(IV) species of Fe-MOF-74. ACS Catalysis. 2015;**5**(6):3287-3291

[58] Chung LW, Sameera WMC, Ramozzi R, Page AJ, Hatanaka M, Petrova GP, et al. The ONIOM method and its applications. Chemical Reviews. 2015;**115**(12):5678-5796. DOI: 10.1021/acscatal.5b00475

[59] Rappé AK, Casewit CJ, Colwell KS, Goddard WA III, Skiff WM. UFF, a full periodic table force field for molecular mechanics and molecular dynamics simulations. Journal of the American Chemical Society. 1992;**114**(25):10024-10035. DOI: 10.1021/ja00051a040

[60] Liao P, Getman RB, Snurr RQ. Optimizing open iron sites in metal-organic frameworks for ethane oxidation: A first-principles study. ACS Applied Materials & Interfaces. 2017;**9**(39):33484-33492. DOI: 10.1021/acsami.7b02195

[61] Impeng S, Siwaipram S, Bureekaew S, Probst M. Ethane C-H bond actIVation on the Fe(IV)-oxo species in a Zn-based cluster of metal-organic frameworks: A density functional theory study. Physical Chemistry Chemical Physics. 2017;**19**(5):3782-3791. DOI: 10.1039/c6cp07771d

[62] Brozek CK, Dincă M. Ti3+–, V2+/3+–, Cr2+/3+–, Mn2+–, and Fe2+–substituted MOF-5 and redox reactivity in Cr-and Fe-MOF-5. Journal of the American Chemical Society. 2013;**135**(34):12886-12891. DOI: 10.1021/ja4064475

[63] Evans JD, Sumby CJ, Doonan CJ. Post-synthetic metalation of metal-organic frameworks. Chemical Society Reviews. 2014;**43**(16):5933-5951. DOI: 10.1039/C4CS00076E

[64] Brozek CK, Dincă M. Cation exchange at the secondary building units of metal-organic frameworks. Chemical Society Reviews. 2014;**43**(16):5456-5467. DOI: 10.1039/C4CS00002A

[65] Rimoldi M, Bernales V, Borycz J, Vjunov A, Gallington LC, Platero-Prats AE, et al. Atomic layer deposition in a metal-organic framework: Synthesis, characterization, and performance of a solid acid. Chemistry of Materials. 2017;**29**(3):1058-1068. DOI: 10.1021/acs.chemmater.6b03880

[66] Liu T, Vermeulen NA, Howarth AJ, Li P, Sarjeant AA, Hupp JT, et al. Adding to the arsenal of zirconium-based metal-organic frameworks: The topology as a platform for solvent-assisted metal incorporation. European Journal of Inorganic Chemistry. 2016;**2016**(27):4349-4352. DOI: 10.1002/ejic.201600627

[67] Bernales V, Ortuño MA, Truhlar DG, Cramer CJ, Gagliardi L. Computational design of functionalized metal–organic framework nodes for catalysis. ACS Central Science. 2017;**4**(1):5-19. DOI: 10.1021/acscentsci.7b00500

[68] Ortuño MA, Bernales V, Gagliardi L, Cramer CJ. Computational study of first-row transition metals supported on MOF NU-1000 for catalytic acceptorless alcohol dehydrogenation. Journal of Physical Chemistry C. 2016;**120**(43):24697-24705. DOI: 10.1021/acs.jpcc.6b06381

[69] Bernales V, League AB, Li Z, Schweitzer NM, Peters AW, Carlson RK, et al. Computationally guided discovery of a catalytic cobalt-decorated metal-organic framework for ethylene dimerization. Journal of Physical Chemistry C. 2016;**120**(41):23576-23583. DOI: 10.1021/acs.jpcc.6b07362

[70] Li Z, Peters AW, Bernales V, Ortuño MA, Schweitzer NM, Destefano MR, et al. Metal-organic framework supported cobalt catalysts for the oxidative dehydrogenation of propane at low temperature. ACS Central Science. 2017;**3**(1):31-38. DOI: 10.1021/acscentsci.6b00290

[71] Bernales V, Yang D, Yu J, Gümüşlü G, Cramer CJ, Gates BC, et al. Molecular rhodium complexes supported on the metal-oxide-like nodes of metal organic frameworks and on Zeolite HY: Catalysts for ethylene hydrogenation and dimerization. ACS Applied Materials & Interfaces. 2017;**9**(39):33511-33520. DOI: 10.1021/acsami.7b03858

The Use of Density Functional Theory to Decipher the Electrochemical Activity of Metal Clathrochelates with Regard to the Hydrogen Evolution Reaction in the Homogeneous Phase

Manuel Antuch and Pierre Millet

Additional information is available at the end of the chapter

http://dx.doi.org/10.5772/intechopen.80267

Abstract

The energetic needs of a rising human population have led to the search for alternative energy sources. A promising route for the large-scale storage of renewable energy is water electrolysis, which is performed with a proton-conducting polymer electrolyte. However, only platinum group metal electrocatalysts have the adequate properties to minimize the overvoltages associated with either hydrogen or oxygen evolution reactions. Alternative materials based on transition metals are scarce, but molecular electrochemistry offers some alternatives. In particular, transition metal clathrochelates exhibit an interesting activity with regard to the hydrogen evolution reaction (HER). However, such complexes form a vast family, and there is a need to implement screening approaches to identify the most performing ones. Theoretical studies on molecular electrocatalysts are adequate for this purpose, since density functional theory (DFT) has a strong predicting capability to provide clues for the improvement of practical devices. This chapter describes the most recent theoretical methods applied to several members of the clathrochelate family. We describe the computation of their common spectroscopic and electrochemical properties. In addition, DFT analysis is used to decipher the multistep reaction mechanism of a model Co clathrochelate with regard to the hydrogen evolution reaction in the homogeneous phase.

Keywords: clathrochelate, cage compounds, DFT, electrochemistry, electrocatalysis, hydrogen evolution reaction

1. Introduction

Clathrochelate is a term used in coordination chemistry to denote the ligands that can encapsulate metal ions. The word is specifically used here as a generic term to designate a family of macrocyclic organometallic compounds that contain tris-glyoximate ligands, forming a cage that is strongly bound to a central metal atom (typically Fe [1] or Co [2]; (**Figure 1**). Interestingly, such clathrochelates form a wide family of compounds since their apical and ribbed substituents may be varied with ease. Electron donor (such as alkyl) or electron-withdrawing substituents (such as halogens) can be used in the structure to adjust the electronic properties of the metallic center. The XRD structure of the cobalt clathrochelates in their resting state (i.e., neutral molecule with the metal ion in a 2+ oxidation state) shows four N atoms of two glyoximate groups closer to Co, typically at a distance between 1.88 to 1.99 Å, and two other N atoms of the third glyoximate group more elongated at a distance of 2.10 to 2.17 Å [2, 3], reflecting a Jahn-Teller distortion. Another structural parameter employed for their description is the dihedral angle (θ) between the atoms B'N'NB (where the apostrophe distinguishes the two apical extremes of the complex), which expresses the rotation between the two N'··N'··N' and N··N··N triangles, as illustrated in **Figure 1**. Both Co—N distances and θ depend on the oxidation state of the molecule, reflecting structural variations occurring upon electron gain or loss.

The relevant property of these clathrochelates that motivates this chapter is their electrocatalytic activity for hydrogen production from protons in the homogeneous phase (i.e., when the compound is dissolved in solution) [2] known as the hydrogen evolution reaction (HER). The versatility of these molecules makes them further suitable for their incorporation in heterogeneous systems for light harvesting [4] or as precursors for the deposition of Co-containing nanoparticles [5], which are also active for the HER.

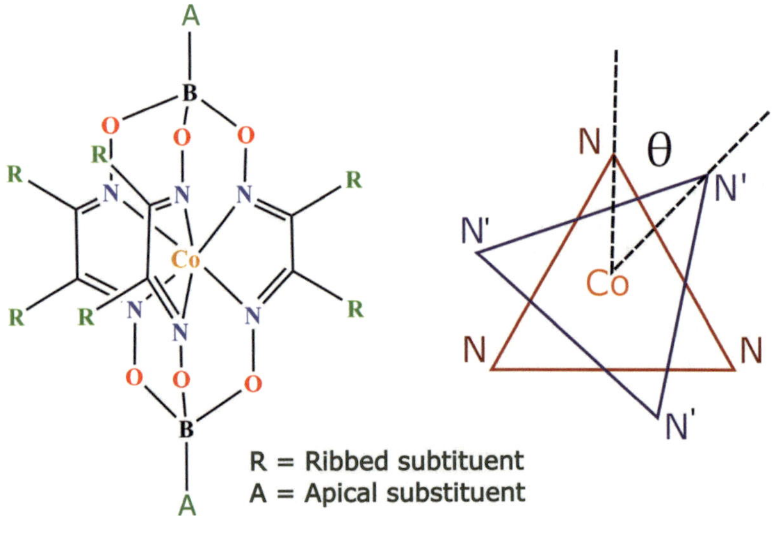

Figure 1. General Lewis structure of a Co clathrochelate. These molecules are versatile since they may bear various apical and ribbed substituents.

Herein, we illustrate the use of DFT to study the electrochemical properties of these clathrochelates with regard to the HER. First, we provide a concise overview of the theoretical methods currently employed to address these complexes and the typical procedures employed to aid and stabilize self-consistent field (SCF) and geometry optimization convergence. Subsequently, we provide examples of the computation of common spectroscopies, such as infrared (IR) or ultraviolet-visible (UV-Vis). The core of this chapter is then devoted to the computation of electron transfer processes and to unravel the HER electrocatalytic mechanism with a model Co clathrochelate. Finally, we provide an overall conclusion of the state of the art regarding the application of DFT methods to clathrochelates, along with conceivable perspectives for future work on the topic.

2. Comments on the selection of the calculation methodology

The choice of the theoretical method is the cornerstone requirement for a successful computation leading to results of predictive value. However, there is no established and unique approach to select the appropriate method for solving a particular problem. Despite this, some general guidelines may be set in order to select the suitable tools.

For example, DFT studies of organic molecules have overwhelmingly been performed with the B3LYP exchange-correlation functional [6], which is currently a standard procedure joined to Pople's basis sets (the family of 6-31G and so on). At the present time, there are no huge problems with the theoretical treatment of relatively small organic molecules. Conversely, organometallic complexes in general, and clathrochelates in particular, are much more challenging, because of the plethora of functionals and basis sets available. Ultimately, it is the comparison with an experimental quantity that tells whether a theoretical method is satisfactory or not. Typically, the initial manner is to skim the literature to find similar systems which have been treated before and to use such methodology. In the case of Co clathrochelates, the literature is dominated by the use of the B3LYP exchange-correlation functional. We have successfully used the B3LYP functional for geometry optimizations, with excellent agreement to experiment [2]; however, all attempts to use other functionals (M06-L, TPSS, BP86, PBE, or PBE0) yielded unreasonable geometries [7]. Regarding the selection of the basis set, the LANL2DZ with effective core potentials for heavy atoms predicts well the geometry of Co clathrochelates; on the other hand, the use of a more complete and more expensive all-electron basis set cc-pVTZ for Co and cc-pVDZ for all other atoms did not improve significantly the results [7].

Regarding geometry optimization convergence, it is reasonable to be generous on the allowed iteration steps; for instance, we have always employed 100 iterations for these kind of problems. Another frequent issue is SCF convergence. We recommend the use of Pulay's direct inversion in the iterative subspace (DIIS) and the damping of the Fock matrix, with the purpose to accelerate and stabilize convergence; alternatively to DIIS, the second-order SCF orbital optimization (SOSCF) method may be used as well. Finally, the computation of SCF may be performed directly or not. This has an impact mostly on the calculation time, and the choice depends on the hardware available for work. In the case of direct SCF, the integrals are calculated each time whenever necessary; on the other hand, if direct SCF is not chosen, integrals are calculated only once and then stored in memory. For computers that have mechanical-based hard disks, it is advisable to use direct SCF, because the calculation of integrals by the processor is generally faster than their search in memory. The time spent for the calculation of clathrochelates may be high, of course depending on the functional, the basis set, and the substituents of the molecule. As an illustrative example, by running a parallelized version of GAMESS [8] on 29 processors, it took 1 week for a complete optimization and Hessian computation of complex $CoBd_3(B—nC_4H_9)_2$ which possesses n-butyl and phenyl substituents in the apical and ribbed positions, respectively, using the B3LYP functional and the LANL2DZ basis set with effective core potentials for Co.

3. Computation of typical spectroscopic properties of a model clathrochelate

A major goal of theoretical chemistry is to be able to predict measurable properties and therefore to provide insights into the behavior of molecular systems. Two of the most ubiquitous techniques available are infrared (IR) and ultraviolet-visible (UV-Vis) spectroscopies, which provide information on the normal modes of vibration (IR) and electronic excitations (UV-Vis). At the end of any geometry optimization, a Hessian calculation should be performed, giving access to the normal modes of the molecule; indeed, all calculated frequencies must be positive in order to verify that the identified stationary point is a true minimum. A typical example of Hessian calculation for the $Co(Cl_2Gm)_3(B—PhF_5)_2$ molecule (where Gm and PhF_5 stand for the glyoximate and the pentafluorophenyl moieties, respectively) is depicted in **Figure 2a**, along with a comparison to the experimental IR spectrum. The calculation was performed with the B3LYP exchange-correlation functional and using the LANL2DZ basis set for all atoms, including effective core potentials for Co and Cl. Note the excellent agreement between experiment and calculation, showing that most signals observed in the experimental spectrum are not pure normal modes but a complex convolution of multiple vibrations that are close in frequency. In particular, the absorption band highlighted with a (1) is essentially composed by a single normal mode. However, such mode is not located in a small part of the molecule but comprises the whole molecular structure, with coupled vibrations from the ligand cage, and the pentafluorophenyl rings, as depicted on the right side of **Figure 2b**.

Another important property of these molecules that may be described is their optical absorption. In order to calculate the UV-Vis characteristics, time-dependent density functional theory (TDDFT) formalism has to be employed. The use of the B3LYP functional for the calculation of these kind of spectra has been shown to be an adequate choice [3].

However, the basis function should be chosen to be more complete than for geometry optimizations and Hessian calculations. In particular, it is desirable to include diffuse functions in the basis set, because the optical excitations may occur to nonbonding, unfilled orbitals which develop farther away from the molecule. Previous predictions of optical spectroscopy of clathrochelates have employed a mixture of basis sets, such as the triple-ζ with a

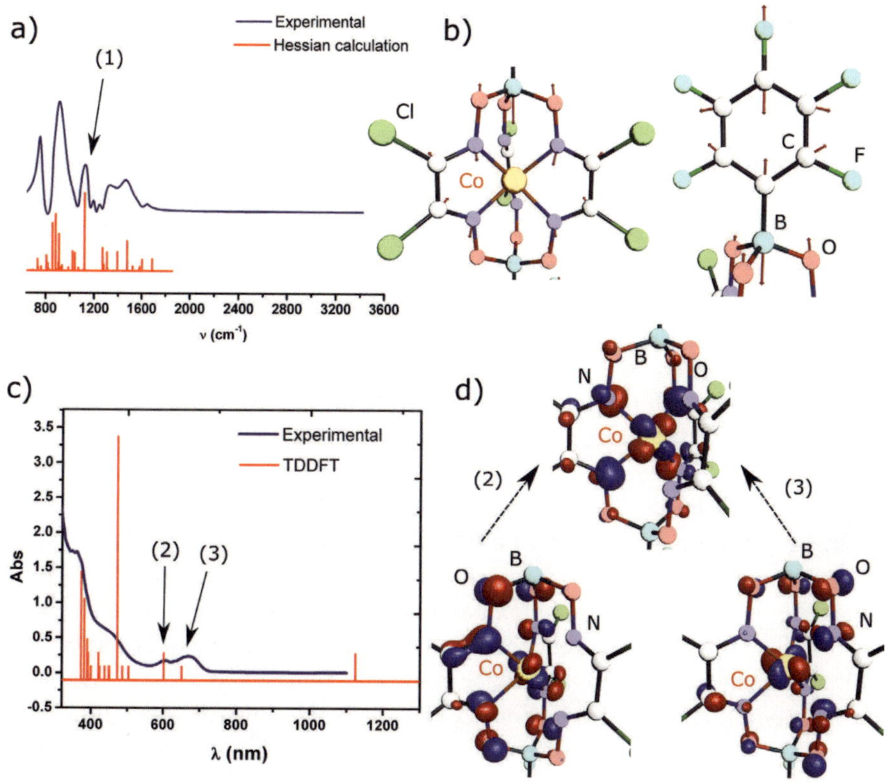

Figure 2. (a) Comparison of the experimental IR spectrum (blue lines) of the $Co(Cl_2Gm)_3(B—PhF_5)_2$ molecule and the hessian calculation (red sticks); (b) illustration of the normal mode highlighted with number (1) in item (a), functional B3LYP, and basis set LANL2DZ with ECP for Co and Cl; (c) experimental UV-Vis spectrum superimposed to the theoretical spectrum calculated by TDDFT. Functional B3LYP and aug-cc-pVTZ basis set for Co and cc-pVDZ for all other atoms.

polarization function (TZVP) for Co and double-ζ split valence for all other atoms [3]. For our calculations (**Figure 2c** and **d**), we have employed the augmented version of the correlation-consistent polarized valence triple-ζ (aug-cc-pVTZ) basis set for Co and the correlation-consistent polarized valence double-ζ (cc-pVDZ) for all other atoms, and we have computed the first 40 excited states. It should be noted that, for the calculation of UV-Vis spectra of usual small organic molecules, it is often sufficient to calculate only four excited states [6], in open contrast to organometallic compounds, where a large number has to be computed in order to adequately reproduce the spectrum. As for IR spectra, these complex molecules exhibit complex UV-Vis spectra, and the assignment of the different contributions is not straightforward because each transition is composed of several contributions from different orbitals. In particular, two transitions occurring in the visible region are highlighted with numbers (2) and (3) in **Figure 2c**. The main orbitals involved in such transition are depicted on the right side of **Figure 2d**. Certainly, these transitions cannot be classified simply into metal-to-ligand charge transfer (MLCT) or ligand-to-metal charge transfer (LMCT), because all orbitals involved in the transition have important contributions from the d orbitals of cobalt and from the glyoxime moieties of the ligand.

4. Computation of electrochemical properties

4.1. Estimation of redox potentials with DFT

DFT calculations can also be used to analyze the redox properties of molecules. Electron transfer reactions may take place between chemical species dissolved in solution or between a molecule in solution and a solid-state electrode or photoelectrode. In particular, electrochemistry is the science that studies matter transformation upon the passage of electric current. Species that gain electrons decrease their oxidation state (and are said to be reduced), while species providing electrons to others increase their oxidation sate (and are said to be oxidized). These oxidation-reduction reactions, which necessarily involve the exchange of electrons but may involve chemical steps as well, are referred with the shorthand term "redox." Certain molecules may exist in two (or more) oxidation states, and these pairs are known as redox couples, represented as O∣R, where O stands for the oxidized form, and R represents the reduced form. The redox reaction $O + e^{-} = R$ occurs at a certain electrode potential ($E°$), which is mainly observable in electrochemistry. The redox potential is important because the more positive it is, the more oxidizing is O; likewise, the more negative it is, the more reducing is R. Essentially, the redox potential is a thermodynamic quantity that allows to predict the feasibility of redox reactions. Absolute redox potentials for molecules are not measurable, and therefore a proper reference electrode is employed. Arbitrarily, the standard hydrogen electrode (SHE) has been adopted as zero in the potential scale [9]. This relative scale is quantitatively related to the absolute energy scale used in physics, i.e., the electron in a vacuum [10].

DFT may provide an estimation of the absolute redox potential of molecules, following two main methodologies. The first method consists in the calculation of a Born-Haber process, by optimizing all relevant species both in gas phase and in the solvent. This procedure is relatively lengthy and therefore shall not receive further attention in this chapter, but the interested reader is referred to the specialized literature [11–14]. Conversely, the full calculation of the Born-Haber process may be bypassed, and the optimization and Hessian calculation of the relevant redox species, i.e., O and R, in the desired solvent may be undertaken directly [7, 15]. The calculation of the absolute redox potential of the species is found according to Eq. (1):

$$E^{\circ}_{abs} = -(\Delta G^{\circ}_{abs}/nF) \tag{1}$$

where F and n represent Faraday's constant and the number of exchanged electrons, respectively, and ΔG^{0}_{abs} is the difference in free energy of the reaction $O + e^{-} = R$. The value of ΔG^{0}_{abs} contains (i) the total electronic energy of the molecule (i.e., electron-electron and nucleus-nucleus repulsions plus electron-nucleus attraction), (ii) zero-point energy (ZPE) corrections, (iii) vibrational corrections to the free energy, and (iv) thermal corrections. We note that the electronic energy of the molecule is obtained by a single-point calculation, but all other thermodynamic corrections are only accessible after a Hessian calculation.

This calculated absolute potential may be referred to an estimated value of the absolute potential of the SHE. Unfortunately, the literature is plenty of different values, for example, 4.2 ± 0.4 V [16] to 4.29 ± 0.02 V [17], 4.05 V [18], and 4.44 V [19], which makes it difficult to assess precision in the calculation, given that all authors provide arguments for their absolute values of the SHE. In order to overcome this drawback, the choice of a reference redox reaction is a safer option. The most commonly used reference reaction is that of the $Fc^{1+}|Fc$ (i.e., ferrocenium|ferrocene) redox couple. When this procedure is used, the optimization and Hessian calculation of both Fc^{1+} and Fc at exactly the same theoretical level than the O|R couple under study are performed (**Figure 3a**). Then, the absolute potential of $Fc^{1+}|Fc$ is obtained using Eq. (1). Obviously, this is a sort of "redox calibration" of the computational method. The theoretical estimation of the redox potential of the O|R couple vs. the $Fc^{1+}|Fc$ couple is just the difference between the absolute potentials these two redox pairs.

Current DFT methods provide estimates of redox potentials with an error of at most 200 mV [11, 12, 20], which is nowadays tacitly considered as the limit for a reasonable estimation. There are several causes for such discrepancies with experimental observations, such as the fact that (i) the exact functional linking the electron density and the electronic energy is not known, (ii) the use of an incomplete basis set, (iii) the consideration of only one conformation of the molecule for the computation, (iv) the treatment of the solvent as an electrostatic continuum, and (v) the harmonic oscillator approximation to estimate thermodynamic quantities. Despite these sources of errors, DFT is powerful in predicting reactivity trends while giving reasonable estimations for observable properties.

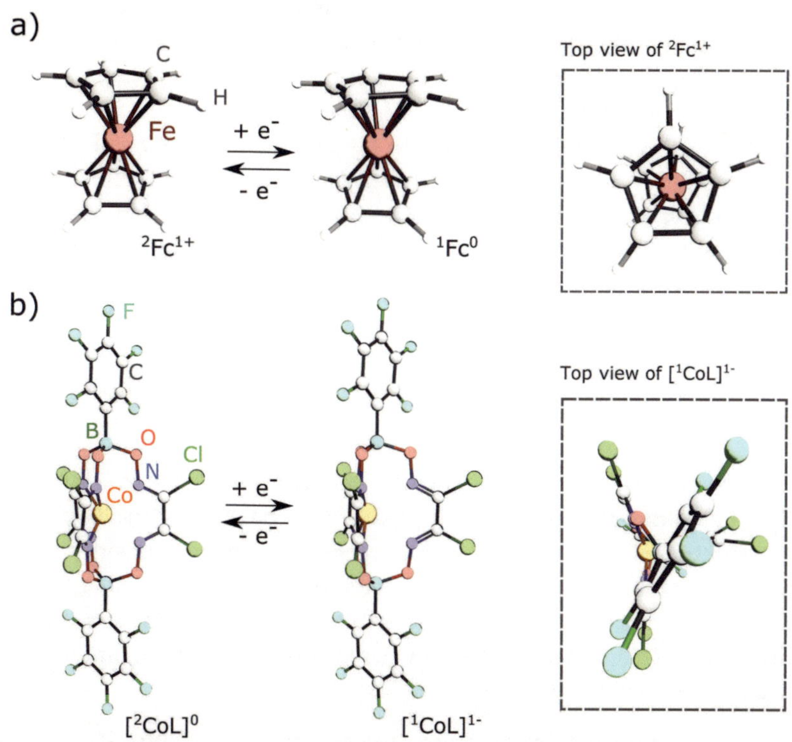

Figure 3. (a) Optimized structures for ferrocene redox chemistry, which is the reference reaction allowing the calculation of redox potentials; (b) optimized geometries of the $Co(Cl_2Gm)_3(B—PhF_5)_2$ in Co(II) and Co(I) oxidation states. Geometries were optimized with the B3LYP functional and the LANL2DZ basis set, with effective core potentials for Co and Cl in dimethylformamide (DMF). The number in the upper left represents the spin multiplicity.

The comparison between theoretical and experimental redox potentials has another subtlety. Indeed, when an aqueous reference electrode, such as the saturated calomel electrode (SCE) or the silver-silver chloride (Ag | AgCl | Cl^-), is used in a nonaqueous environment, there is an inherent liquid junction potential (LJP) [21] which depends on electrolyte composition and biases the measurement. Hence, in order to compare theoretical results with experimental values, the first issue to pay attention to is that an aqueous reference had not been used to make the measurement in a nonaqueous environment. If an aqueous reference has been used in a nonaqueous solvent the Co atom moves to one side, it is still possible to make the comparison with theory, by finding a literature value for the Fc^{1+} | Fc couple and converting the potential scale of the aqueous reference (which was used in a nonaqueous solvent) to the Fc^{1+} | Fc scale. Literature values for Fc^{1+} | Fc have been reported as +0.50 V vs. Ag | AgCl | Cl^- [22] or + 0.46 V vs. SCE [2], comprising the LJP.

Finally, it is worth to stress that the redox potential and the energy of molecular orbitals are different things and cannot be equated. Certainly, when representing electron transfer reactions, the educational literature often depicts electrons as being transferred from frontier orbitals [23]. However, such representations might induce to think that the energy of orbitals is somehow linked to redox potentials. As a matter of fact, molecular orbitals are mathematical functions that describe the electron motion in a molecule and are certainly not observable [24], in contrast to the redox potential which is an observable thermodynamic quantity. Despite this, the energetic trend in orbital energies may be commonly paralleled to the oxidizing/reducing power of molecules. For example, with all other things being equal, the more electronegative the substituents in the ligand of a complex, the lower the energy of the molecular orbitals. Hence, the complex will be more likely to gain an electron in the lowest unoccupied molecular orbital (LUMO) than to give it from the highest occupied molecular orbital (HOMO); this is equivalent to say that the complex shall be a better oxidant and its redox potential should be more positive. This kind of trends, linking the energy of molecular orbitals and redox potentials, has been presented several times in the literature [25, 26]. Though we have shown that the Co^{III} | Co^{II} couple in a series of clathrochelates complies with it, the Co^{II} | Co^{I} redox pair does not follow the orbital trends. Further details may be found in Ref. [2].

4.2. Mechanistic investigation of the electrocatalytic hydrogen evolution reaction and ligand noninnocence

The origin of the catalytic activity of metal-containing clathrochelates is currently under debate [1]. It is likely that the role of the ligand is essential to electrocatalysis, since it may stabilize low oxidation states in the metal to enable electron transfer or may participate in protonation reactions that may help bring protons together and evolve H_2. That is why the term "ligand non-innocence" has been coined to refer to organometallic complexes in which the organic ligand has a major participation in its redox behavior to account for high reactivity involving small species, such as CO_2 or H^+ [27–29].

For example, upon the one-electron reduction process depicted in **Figure 3b**, we have predicted interesting structural modifications within the molecule. The asymmetry inside the Co—N cage is calculated to be reinforced, with four Co—N distances being at around 1.88 Å and the two other N lying farther apart at 2.45 Å; consequently, the Co atom moves to one side of the cage, thus decoordinating from two N atoms and resulting in a square planar geometry for the reduced complex; furthermore, there is a slight shortening of N—C(sp_2) bonds from 1.30 to 1.28 Å.

Ligand non-innocence in the case of clathrochelates may be further investigated because their chemical structure invites to think on multiple sites as putative candidates for protonation. Previous theoretical studies have addressed only protonation on the iminic C(sp_2) [30], but we have extended such approach by further exploring other protonation sites. Indeed, the basic character of N and the wide evidence of metal hydrides as intermediates in the HER have led us to explore such alternative protonation sites.

Until now, the most complete theoretical study of hydrogen evolution involving clathrochelates is found in Ref. [7]. The work was performed on a clathrochelate having chlorine and methyl moieties as the ribbed and apical substituents, respectively. Henceforward, this epigraph presents those results in a concise manner.

Optimized structures for three protonation possibilities are illustrated in **Figure 4**; the investigated protonation sites are highlighted with a red square (C), blue circle (N), and an orange triangle (Co). The insets on the right side of **Figure 4** show a magnification of the coordination position of H^+. Protonation of the iminic C(sp_2) brings a hybridization change to (sp_3) with the concomitant change in the geometry of such C from planar to tetrahedral. Protonation of N has a similar effect on the local geometry of the ligand, which is modified from planar to tetrahedral as well. Direct protonation of Co might seem unlikely at the first sight because the metal appears buried inside the organic cage. However, the movement of Co inside the

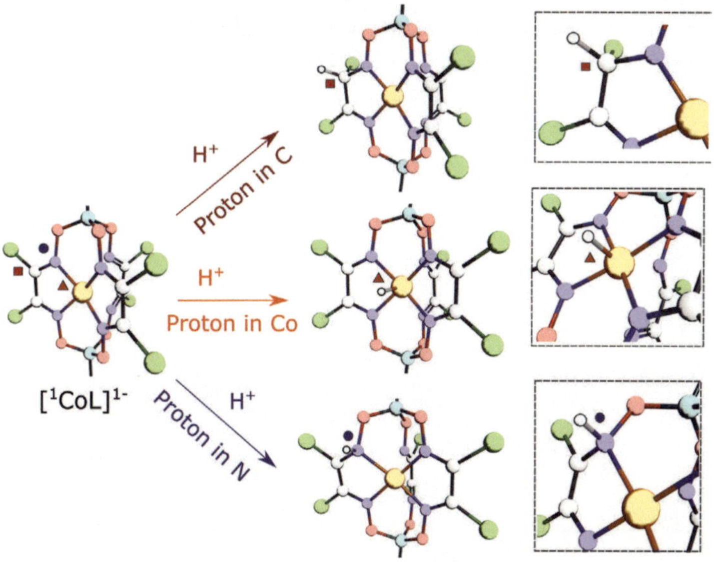

Figure 4. Magnification of the organic cage surrounding the Co-containing hexachloride clathrochelate $Co(Cl_2Gm)_3(B—CH_3)_2$ and the different protonation sites available. Geometries were optimized with the B3LYP functional and the LANL2DZ basis set, with effective core potentials for Co and Cl in dimethylformamide (DMF).

ligand cage upon the one-electron reductive activation and the associated decoordination from two N makes the metal accessible for protonation as well. The calculated X-H distances (where X stands for Co, N, or Csp_2) for these three species are 1.09 Å for Csp_2-H, 1.45 Å for Co—H, and 1.03 Å for N—H. Such bond distances show that proton binding to N or C results in the formation of a strong bond, in contrast to the binding to the metal center, where a much more elongated Co(III)—H bond is created, indicating a weaker binding and therefore a more labile H^+.

An energetic analysis of this initial part of the mechanism is illustrated in **Figure 5a**. Indeed, there is an energetic gain upon the reductive activation of the $[^2CoL]^0$ to yield the catalytically active $[CoL]^{1-}$ intermediate. Then, protonation of N or $C(sp_2)$ is likely, but the most favored protonation site was predicted to be the Co central ion.

We have also addressed the possibility of double protonation in two adjacent N; in a $C(sp_2)$ and a N; in two adjacent $C(sp_2)$; in Co and N; and in Co, for both high-spin and low-spin species. However, most optimized structures were largely distorted due to heavy structural changes caused by the sp_2 to sp_3 hybridization changes that should occur after protonation. As a consequence, the calculated energy of doubly protonated intermediates was prohibitively high, and these species are unlikely to be formed. The most stable doubly protonated species was the one with protons coordinated in two adjacent N atoms. However, it seems unlikely that the mechanism of hydrogen evolution involved initially a mono-protonated species that is a cobalt hydride and that the subsequent species in the mechanism is a double-protonated intermediate with protons in adjacent N atoms.

The Co(III)—H intermediate is interesting and deserved closer attention. In particular, such species may undergo a further one-electron reduction to yield Co(II)—H, which may also have a role in the mechanism. Certainly, we have predicted that such reduction brings an important energetic gain, but the search for doubly protonated intermediates and the search for a transition state to explain hydrogen evolution were unsuccessful. We were able to locate a transition state of the η^2—Co(III)—(HH) type, but again this was far too energetic to be a plausible candidate to explain hydrogen evolution (**Figure 6b**).

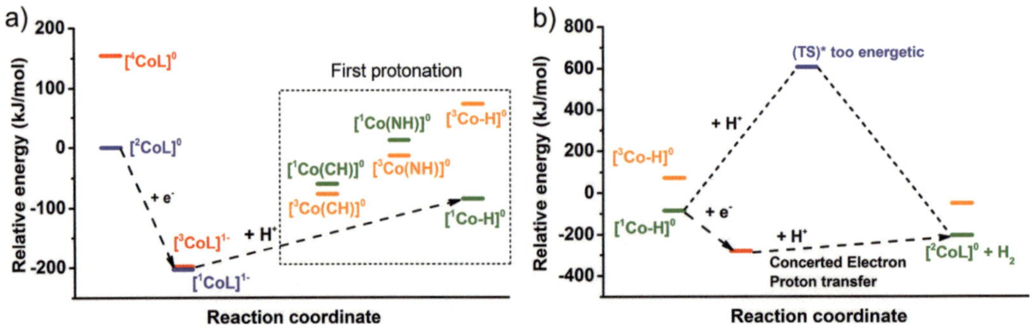

Figure 5. (a) Energetic diagram of the first electron transfer to produce the catalytically active $[Co^I]^{1-}$ species and its further protonation at different sites of the ligand, as depicted in figure X; (b) representation of the second part of the mechanism, showing the unfeasibility of hydrogen evolution through a transition state of high energy and the alternative possibility of concerted proton-electron transfer.

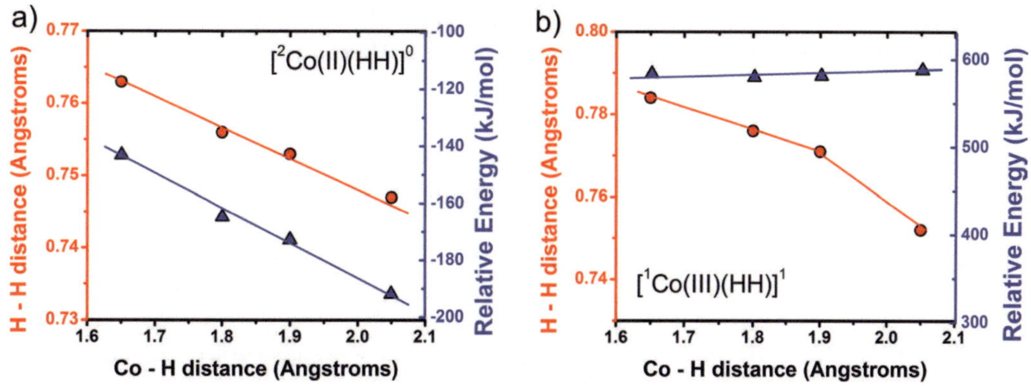

Figure 6. Main results from potential energy surface scans showing the relation between the Co—(HH) and the H—H bond distance along with the relative energy involved during the separation of H_2 from the complex, (a) Co(II) and (b) Co(III). These calculations were performed in the complex $Co(Cl_2Gm)_3(B—CH_3)_2$ at the B3LYP level and the LANL2DZ basis set, with effective core potentials for Co and Cl. The energy was further refined with the more complete all-electron basis set cc-pVTZ for Co and cc-pVDZ for all other atoms in dimethylformamide (DMF). The zero in the energy scale is the [^{2}CoL]^0 species.

Considering that the hydrogen production step is of key importance, we have devoted wide efforts to explain how it happens. When the calculations on a mechanism yield high energy intermediates, it is possible that a concerted electron-proton transfer process occurred, in order to avoid energy-demanding paths [15, 20]. Thus, potential energy surface (PES) scans were performed with the Co(II)—(HH) and Co(III)—HH) species at different (fixed) Co—H distances, while the rest of the structure was allowed to optimize freely. These results are summarized in **Figure 6a** and **b**. Interestingly, in the case of Co(II)—(HH), the increase in the Co—H bond distance was linked to a decrease in the H—H distance, approaching the equilibrium bond distance for H_2, which is 0.74 Å. At the same time, the energy of the system gradually decreased to approach the final energy level of the reaction (observe the blue scale on the right side of **Figure 6a**), which is depicted in **Figure 5b** as [^{2}CoL]^0 + H_2. These calculations show a nice picture of two protons that come together in the nearby of Co(II) and bind to one another to form H_2 while going away from the metal, with the concomitant decrease in total energy. Conversely, the same calculations on the Co(III) species gave the same bond-breaking (Co—H) bond-formation (H—H) sequence as just described for Co(II). However, at the same Co—H distance, the H—H separation was consistently higher in the case of Co(III), thus indicating that Co(III) was less effective than Co(II) to promote the reduction and binding of two protons. Moreover, the total energy of the system for the Co(III) ion did not decrease smoothly as in the case of Co(II) (**Figure 6b**). Instead, it was excessively high, and therefore such intermediate was definitively ruled out as a possibility to explain the catalytic activity of this clathrochelate.

5. Conclusions and perspectives

The clathrochelate family has received attention due to its multiple applications. In this chapter, we have illustrated the use of density functional theory to describe several properties of clathrochelates. We began the presentation by providing tips on the choice of the calculation methodology to address these complex molecules. Then, the spectroscopic properties (IR and UV-Vis) have been reproduced. The possibility to accurately predict spectroscopic properties gives an adequate degree of confidence on the theoretical procedures employed. Subsequently, we have illustrated the procedures to predict the redox potentials of molecules, and we have highlighted some structural changes underwent during the electron transfer process. This chapter has introduced the concept of "ligand non-innocence," which is relevant in complex electrocatalytic reactions involving complexes of transition metals, where the ligand has a major role to play, not only to stabilize unusually low oxidation states but also to act cooperatively to bind small species, such as H^+. Finally, a methodology has been presented to address the complex mechanism of hydrogen evolution. Certainly, the hydrogen production step was particularly intriguing for the $Co(Cl_2Gm)_3(B{-}CH_3)_2$ complex, given that neither doubly protonated intermediates nor plausible transition states could be found. Therefore, a concerted proton-electron transfer step was envisaged, with the Co(II)—(HH) hydride intermediate playing the most relevant role. The calculation of potential energy surface scans fixing the Co(II)—(HH) bond distance showed that the simultaneous elongation of Co—H bond leads to the decrease in H—H distance and the smooth decrease in energy, which nicely explained hydrogen evolution from this complex in the homogeneous phase. Overall, results reported in this chapter could contribute to stimulate theoretical and experimentalist chemists to explore the influence of different substituents in the organic cage on the HER activity, thereby providing new insights to researchers that try to develop and optimize alternative electrocatalysts for the HER. Clearly, the formal theoretical understanding of clathrochelate's electrochemistry is still in its early stages, and the missing piece in this scientific challenge is to apply the theoretical methodology outlined in this chapter to other clathrochelates in order to justify the distinct reactivity observed experimentally among different complexes. Moreover, the approach proposed here should provide guidelines to synthetic chemists in the search for more active compounds. Thus, we foresee that the future theoretical research shall be performed in such direction. This is a stimulating moment to conduct research aiming to find novel and efficient catalysts for hydrogen production, and transition metal clathrochelates are perhaps the family of complexes that has received the least attention from the chemical community.

Acknowledgements

Manuel Antuch is thankful to the French ministerial scholarship and the Initiative d'Excellence (IDEX) program for funding.

Conflict of interest

The authors do not have conflict of interest.

Author details

Manuel Antuch* and Pierre Millet

*Address all correspondence to: manuel.antuch-cubillas@u-psud.fr

Research and Innovation Team in Electrochemistry for Energy (ERIEE), Institute of Molecular and Materials Chemistry (ICMMO), UMR CNRS 8182, Paris-Sud University, Paris-Saclay University, Orsay, France

References

[1] Voloshin YZ, Novikov VV, Nelyubina YV, Belov AS, Roitershtein DM, Savitsky A, Mokhir A, Miehlich ME, Meyer K. Synthesis and characterization of an Fe (I) cage complex with high stability towards strong H-acids. Chemical Communications. 2018;**54**: 3436-3439. DOI: 10.1039/c7cc09611a

[2] Antuch M, Ranjbari A, Grigoriev S, Al-Cheikh J, Villagrá A, Assaud L, Voloshin YA, Millet P. Effect of the ligand framework of cobalt clathrochelates on hydrogen evolution electrocatalysis: Electrochemical, spectroscopic and density functional theory analyses. Electrochimica Acta. 2017;**245**:1065-1074. DOI: 10.1016/j.electacta.2017.03.005

[3] Nguyen MTD, Charlot MF, Aukauloo A. Structural, electronic, and theoretical description of a series of cobalt clathrochelate complexes in the Co(III), Co(II) and Co(I) oxidation states. The Journal of Physical Chemistry. A. 2011;**115**:911-922. DOI: 10.1021/jp1103643

[4] Antuch M, Millet P, Iwase A, Kudo A, Grigoriev SA, Voloshin YZ. Characterization of Rh: $SrTiO_3$ photoelectrodes surface-modified with a cobalt clathrochelate and their application to the hydrogen evolution reaction. Electrochimica Acta. 2017;**258**:255-265. DOI: 10.1016/j.electacta.2017.10.018

[5] Anxolabéhère-Mallart E, Costentin C, Fournier M, Nowak S, Robert M, Savéant JM. Boron-capped tris(glyoximato) cobalt clathrochelate as a precursor for the electrodeposition of nanoparticles catalyzing H2 evolution in water. Journal of the American Chemical Society. 2012;**134**:6104-6107. DOI: 10.1021/ja301134e

[6] de la Vega-Hernández K, Antuch M, Cuesta-Rubio O, Núñez-Figueredo Y, Pardo-Andreu GL. Discerning the antioxidant mechanism of rapanone : A naturally occurring benzoquinone with iron complexing and radical scavenging activities. Journal of Inorganic Biochemistry. 2017;**170**:134-147. DOI: 10.1016/j.jinorgbio.2017.02.019

[7] Antuch M, Millet P. Approach to the mechanism of hydrogen evolution electrocatalyzed by a model Co clathrochelate: A theoretical study by density functional theory. Chemphyschem. 2018. DOI: 10.1002/cphc.201800383

[8] Schmidt MW, Baldridge KK, Boatz JA, Elbert ST, Gordon MS, Jensen JH, Koseki S, Matsunaga N, Nguyen KA, Su S, Windus TL, Dupuis M, Montgomery JA. General atomic and molecular electronic structure system. Journal of Computational Chemistry. 1993;**14**:1347-1363. DOI: 10.1002/jcc.540141112

[9] Bard AJ, Faulkner LR. Electrochemical Methods Fundamentals and Applications. 2nd ed. John Wiley & Sons, Inc.; 2001

[10] Archer MD, Nozik AJ. Nanostructured and photoelectrochemical systems for solar photon conversion. In: Series on Photoconversion of Solar Energy. UK: Imperial College Press; 2008;**3**:28-29

[11] Baik M, Friesner RA. Computing redox potentials in solution: Density functional theory as a tool for rational design of redox agents. The Journal of Physical Chemistry. A. 2002;**106**:7407-7412

[12] Fu Y, Liu L, Yu H, Wang Y, Guo Q. Quantum-chemical predictions of absolute standard redox potentials of diverse organic molecules and free radicals in acetonitrile. JACS. 2005:7227-7234

[13] Lewis A, Bumpus JA, Truhlar DG, Cramer CJ. Molecular modeling of environmentally important processes: Reduction potentials. Journal of Chemical Education. 2004;**81**:596-604. DOI: 10.1021/ed081p596

[14] Benassi R, Ferrarini P, Fontanesi C, Benedetti L, Paolucci F. Calculated electron affinities and redox E0 values of polypyridinic derivatives. Journal of Electroanalytical Chemistry. 2004;**564**:231-237. DOI: 10.1016/j.jelechem.2003.10.029

[15] Solis BH, Hammes-Schiffer S. Proton-coupled Electron transfer in molecular Electrocatalysis : Theoretical methods and design principles. Inorganic Chemistry. 2014;**53**: 6427-6443

[16] Donald WA, Leib RD, O'Brien JT, Bush MF, Williams ER. Absolute standard hydrogen electrode potential measured by reduction of aqueous nanodrops in the gas phase. Journal of the American Chemical Society. 2008;**130**:3371-3381. DOI: 10.1021/ja073946i

[17] Donald WA, Leib RD, O'Brien JT, Williams ER. Directly relating gas-phase cluster measurements to solution-phase hydrolysis, the absolute standard hydrogen electrode potential, and the absolute proton solvation energy. Chemistry - A European Journal. 2009;**15**:5926-5934

[18] Donald WA, Leib RD, Demireva M, Brien JTO, Prell JS, Williams ER. Directly relating reduction energies of gaseous Eu(H_2O)n3+, n = 55-140, to aqueous solution: The absolute SHE potential and real proton solvation energy. JACS. 2009;**131**:13328-13337

[19] Trasatti S. The absolute electrode potential: An explanatory note. Pure and Applied Chemistry. 1986;**58**:955-966

[20] Song J, Klein EL, Neese F, Ye S. The mechanism of homogeneous CO_2 reduction by Ni(cyclam): Product selectivity, concerted proton–Electron transfer and C–O bond cleavage. Inorganic Chemistry. 2014;**53**:7500-7507

[21] Kolthoff IM, Thomas G. Electrode potentials in acetonitrile. estimation of the liquid junction potential between acetonitrile solutions and the aqueous saturated calomel electrode. The Journal of Physical Chemistry. 1966;**69**:3049-3058

[22] Fogeron T, Porcher JP, Gomez-Mingot M, Todorova T, Chamoreau L-M, Mellot-Draznieks C, Li Y, Fontecave M. A cobalt complex with a bioinspired molybdopterin-like ligand: a catalyst for hydrogen evolution. Dalton Transactions. 2016;**2**:14754-14763. DOI: 10.1039/C6DT01824F

[23] Elgrishi N, Rountree KJ, McCarthy BD, Rountree ES, Eisenhart TT, Dempsey JL. A practical Beginner's guide to cyclic voltammetry. Journal of Chemical Education. 2018;**95**:197-206. DOI: 10.1021/acs.jchemed.7b00361

[24] Autschbach J. Orbitals: Some fiction and some facts. Journal of Chemical Education. 2012;**89**:1032-1040. DOI: 10.1021/ed200673w

[25] Xuea Z-M, Ding Y-Z, Chen C-H. A DFT study of electronic structures, energies, and molecular properties of lithium bis[croconato]borate and its derivatives. Electrochimica Acta. 2007;**53**:990-997

[26] Al-Anber M, Milde B, Alhalasah W, Lang H, Holze R. Electrochemical and DFT-studies of substituted thiophenes. Electrochimica Acta. 2008;**53**:6038-6047

[27] Panetier JA, Letko CS, Tilley TD, Head-Gordon M. Computational characterization of redox non-innocence in cobalt-bis(diaryldithiolene)-catalyzed proton reduction. Journal of Chemical Theory and Computation. 2016;**12**:223-230. DOI: 10.1021/acs.jctc.5b00968

[28] Lacy DC, Mccrory CCL, Peters JC. Studies of cobalt-mediated electrocatalytic CO_2 reduction using a redox-active ligand. Inorganic Chemistry. 2014;**53**:2-10. DOI: 10.1021/ic403122j

[29] Grelaud G, Gauthier N, Luo Y, Paul F, Fabre B, Barrière F, Ababou-Girard S, Roisnel T, Humphrey MG. Redox-active molecular wires derived from dinuclear ferrocenyl/ruthenium(II) alkynyl complexes: Covalent attachment to hydrogen-terminated silicon surfaces. Journal of Physical Chemistry C. 2014;**118**:3680-3695. DOI: 10.1021/jp412498t

[30] El Ghachtouli S, Fournier M, Cherdo S, Guillot R, Charlot M-F, Anxolabéhère-Mallart E, Robert M, Aukauloo A. Monometallic cobalt – trisglyoximato complexes as precatalysts for catalytic H_2 evolution in water. Journal of Physical Chemistry C. 2013;**117**:17073-17077

Chapter 5

The Application of Periodic Density Functional Theory to the Study of Uranyl-Containing Materials: Thermodynamic Properties and Stability

Francisco Colmenero Ruiz

Additional information is available at the end of the chapter

http://dx.doi.org/10.5772/intechopen.79558

Abstract

With the advent of increased computer capacities, improved computational resources, and easier access to large-scale computer facilities, the use of density functional theory methods has become nowadays a frequently used and highly successful approach for the research of solid-state materials. However, the study of solid materials containing heavy elements as lanthanide and actinide elements is very complex due to the large size of these atoms and the requirement of including relativistic effects. These features impose the availability of large computational resources and the use of high quality relativistic pseudopotentials for the description of the electrons localized in the inner shells of these atoms. The important case of the description of uranyl-containing materials and their properties has been faced recently. The study of these materials is very important in the energetic and environmental disciplines. Uranyl-containing materials are fundamental components of the paragenetic sequence of secondary phases that results from the weathering of uraninite ore deposits and are also prominent phases appearing from the alteration of the spent nuclear fuel. The development of a new norm-conserving relativistic pseudopotential for uranium, the use of energy density functionals specific for solids, and the inclusion of empirical dispersion corrections for describing the long-range interactions present in the structures of these materials have allowed the study of the properties of these materials with an unprecedented accuracy level. This feature is very relevant because these methods provide a safe, accurate, and cheap manner of obtaining these properties for uranium-containing materials which are highly radiotoxic, and their experimental studies demand a careful handling of the samples used. In this work, the results of recent applications of theoretical solid state methods based on density functional theory using plane waves and pseudopotentials to the determination of the thermodynamic properties and stability of uranyl-containing materials are reviewed. The knowledge of these thermodynamic properties is indispensable to model the dynamical behavior of nuclear materials under diverse geochemical conditions. The theoretical methods provide a profound understanding of the thermodynamic stability of these mineral phases and represent a powerful predictive tool to determine their thermodynamic properties.

Keywords: uranyl-containing minerals, spent nuclear fuel, density functional theory, thermodynamic stability, heat capacities, entropies, enthalpies, Gibbs free energies, thermodynamic properties of formation, Gibbs free energies of reaction

1. Introduction

The basis of thermodynamic theory [1] is known since the end of the nineteenth century, and the fundamental developments carried out in the twentieth century have established this theory as a self-contained system of knowledge. The thermodynamic calculations are mainly used for the description of the changes of state associated with the transfer of matter and energy and are an indispensable part of technical and scientific investigations in various fields such as chemistry, metallurgy, chemical engineering, and the energy and environmental technologies. The reliability of the results of thermochemical calculations depends, in the first instance, on the accuracy of the thermochemical data used and the inclusion of the most important species involved, which are often quite numerous if one desires to obtain a fair description of real systems.

There are different sources of data available for establishing thermodynamic information such as calorimetric and solubility measurements [2, 3], phase equilibrium data [4], experimental data on solid solutions [5], and heat capacities and entropies estimated from lattice vibrational models [6–8]. Reasonably complete sets of basic data needed for the calculation of thermochemical functions are available only for a relatively small number of substances [9], and frequently these sets must be completed with empirical data of lower accuracy obtained by analyzing values from different sources as well as by performing estimations before they can be used for actual calculations.

The data listed in the thermochemical tables generally form databases [10–18, 4]. An internally consistent database is one which permits the computation of phase equilibrium relations as established by experimental studies, and it is at the same time compatible with calorimetric and other measurements of thermochemical properties of the phases involved. The generation of such databases is very complex due to the large uncertainties associated with phase equilibrium studies at high temperature and pressure and because many experiments may be irreversible. The CALPHAD method (CALculation of PHAse Diagrams) [19, 20] may be used in the derivation and assessment of this kind of databases [21].

These databases can be used to carry out thermodynamic multi-component, multi-phase, multi-reaction equilibrium calculations on systems made up of any of the substances present in the database. These calculations are best performed by adopting the method of minimization of the total Gibbs free energy of the closed system [22–28]. A detailed description of the principles and techniques used in the computation of equilibrium assemblages of a closed system can be found in the work of Smith and Missen [24]. Thus, these databases are the basis for the software packages for the calculation of equilibria in multi-component systems [12, 13, 18, 19, 22, 23, 29–32]. Examples of successful applications of thermodynamic techniques to the computation of equilibrium phase assemblages in geological and planetary systems have been reported [33–36].

In the field of nuclear technology, the thermodynamic information is indispensable in order to predict the chemical behavior and dynamics of nuclear materials under diverse environmental conditions. The knowledge of precise thermodynamic data is fundamental for the development of geochemical models for nuclear fuel degradation, to evaluate the origin and evolution of uranium ore bodies, in developing programs for the solution mining of uranium deposits or mine dumps, in the study of spent nuclear fuel (SNF) radioactive waste and in the containment of such waste, and may also be of importance in reactions within breeder reactors [37–41].

The behavior of a deep geologic repository (DGR) of high level radioactive waste (HLWR) will depend mainly on the interaction between the SNF and their surroundings. The hydration and corrosion of the SNF under oxidizing conditions will result in the dissolution of the uranium dioxide composing the SNF matrix and the formation of uranyl secondary mineral phases [42–54]. Therefore, the formation and stability of uranyl minerals will determine the release of U(VI) and other actinide elements from the HLWR container and subsequently from the repository to the biosphere [55–70].

The stabilities and dissolution rates of uranyl minerals are functions of the solution composition, temperature, and local conditions (mainly pH and electrochemical potential), and their prediction requires the knowledge of the Gibbs free energy, enthalpy, and entropy thermodynamic functions of formation for each phase of interest and their variation with temperature. The simulation of the release of uranium from DGRs under oxidizing conditions and the mobility of uranium in the environment can be only performed if the thermodynamic properties of the secondary uranyl minerals that may form in the DGR are available. Consequently, the knowledge of the thermodynamic parameters is crucial for predicting DGR performance [71–73]. However, reliable temperature-dependent thermodynamic data are completely lacking, except for the simplest uranyl-containing materials. Therefore, the development of a complete thermodynamic database for these minerals is mandatory.

The rapid development of the nuclear technology strongly encouraged the research on the field of thermodynamics of nuclear materials and the development of nuclear thermodynamic databases [74, 75]. The great significance of the thermodynamic information of materials containing uranium and related elements in the assessment of the safety of DGRs is reflected by the large number of recent experimental works leading to large reviews and updates of thermodynamic properties of materials [72–77]. Among these studies, we may remark the recent experimental measurements by means of solubility and calorimetry techniques of the thermodynamic properties of uranyl peroxide hydrates [51–54], uranyl carbonate minerals [78], uranyl phosphate and orthophosphate minerals [79], and uranyl silicates [71, 80–84].

Despite of the fast progress in the generation of the nuclear thermodynamic database, there are many uranium-containing materials for which the corresponding data are unreliable due to the large experimental uncertainties [38]. A large amount of effort has been dedicated to the assignment and correction of these uncertainties by means of the implementation of new statistical methods for hypothesis testing and the improvement of the techniques used for measuring the thermodynamic properties of these systems [73]. The need to make available a comprehensive, internationally recognized and quality assured chemical thermodynamic database that meets the modeling requirements for the safety assessment of radioactive waste disposal systems prompted the Radioactive Waste Management Committee (RWMC) of the Organization for Economic Co-operation and Development (OECD) Nuclear Energy Agency (NEA) to launch the Thermochemical Database Project (NEA TDB). The RWMC assigned a high priority to the critical review of relevant chemical thermodynamic data of inorganic species, actinide compounds, and fission products [72, 73]. Besides, the range of conditions of temperature and pressure for which the thermodynamic parameters are available for most nuclear materials is rather limited.

While the knowledge of the temperature dependence of these properties for anhydrous uranium oxides is very complete [73–75], the corresponding data for the secondary phases which arise from alteration of SNF under final DGR conditions are surprisingly scarce. For these secondary phases, the thermodynamic parameters are known only for the standard state (298.15 K and 1 bar). The lack of temperature-dependent information for these phases rules out the possibility of performing reliable thermodynamic modeling studies for the performance assessment of DGRs for SNF. Because the corresponding information is available for anhydrous species, thermodynamic computations have been performed for the uranium-oxygen and sodium-uranium-oxygen system [85–95]. A detailed analysis of previous studies [41, 96–105] suggests that first principles methodology is an excellent complement to experimental methodology for determining the thermodynamic functions of these materials.

In this work, the computation of the thermodynamic properties of a large set of uranyl-containing materials is reviewed [99–105]. This set includes the uranyl peroxide studtite [$(UO_2)O_2{\cdot}4H_2O$] and metastudtite [$(UO_2)O_2{\cdot}2H_2O$], the uranyl hydroxide dehydrated schoepite [$UO_2(OH)_2$], the uranyl oxyhydroxide schoepite [$(UO_2)_8O_2(OH)_{12}{\cdot}12H_2O$], metaschoepite [$(UO_2)_8O_2(OH)_{12}{\cdot}10H_2O$] and becquerelite [$Ca(UO_2)_6O_4(OH)_6{\cdot}8H_2O$], the uranyl silicate soddyite [$(UO_2)_2(SiO_4){\cdot}2H_2O$], the uranyl carbonate rutherfordine [UO_2CO_3], and gamma uranium trioxide [γ-UO_3]. The first eight materials have been identified to be basic components of the paragenetic sequence of secondary phases arising from the alteration of uraninite ore deposits and corrosion of SNF under the final DGR conditions [42–54], and gamma uranium trioxide is the main oxide of hexavalent uranium [100, 106, 107]. Uranyl peroxides appear in the earlier stages of this paragenetic sequence [50–54, 108–111] due to the production of hydrogen peroxide and other oxidants resulting from the radiolysis of water due to the ionizing radiation of the SNF. The uranyl oxyhydroxides also begin to appear soon from the alteration of uranium dioxide [42–48]. Studtite, schoepite, metaschoepite, and becquerelite phases have been observed as alteration products of spent fuel in cooling basins at the Hanford Site (Washington) [112–116] and on Chernobyl "lava" formed during the nuclear accident that occurred in 1986 [117]. The next mineral phases appearing in this sequence are uranyl silicates and, less frequently, uranyl phosphates [42–48]. Uranyl carbonates may precipitate where the evaporation is significant, and the carbon dioxide partial pressure is large [49, 118]. The main ingredients of this paragenetic sequence were inferred by Frondel already in 1956 [42, 43]. The sequence is still widely accepted today [44–48, 118, 119]. However, our knowledge of this sequence is only qualitative, and the performance assessment of the DGRs for HLRW and many other applications in nuclear technology require its quantitative specification.

The crystal structures of these materials [100, 102, 105, 120–123] were successfully determined by means of density functional theory using plane waves and pseudopotentials [124]. A new norm-conserving relativistic pseudopotential specific for uranium atom was generated from first principles with this purpose [101, 121]. Then, using these optimized structures, the thermodynamic properties of these materials were determined including specific heats, entropies, enthalpies, and Gibbs free energies [99–105]. The computed thermodynamic properties were combined with those of the corresponding elements in order to determine the enthalpy and free energy of formation of these materials and its variation with temperature [102, 104, 105]. The methods used in the computation of these thermodynamic functions are briefly described in Section 2. Additionally, the calculation of the Gibbs free energies of reaction and associated reaction constants is also described in this section. The main results obtained are described in Section 3, including a study of the thermodynamic stability of the secondary phases of SNF [102, 104, 105]. Finally, the main conclusions are given in Section 4.

2. Methods

2.1. Computational methodology

The generalized gradient approximation (GGA) together with PBE functional [125] supplemented with Grimme empirical dispersion correction [126] was used to study the uranyl-containing materials such as studtite, metastudtite, dehydrated schoepite, schoepite, metaschoepite, becquerelite, soddyite, rutherfordine, and gamma uranium trioxide [99–105, 120–123]. The introduction of dispersion corrections improved significantly the computed structural, vibrational, mechanic and thermodynamic properties of studtite, metastudtite, dehydrated schoepite, schoepite, metaschoepite, becquerelite, and soddyite as a consequence of the better description of the hydrogen bonding present in the crystal structures of these materials. However, for the case of rutherfordine and gamma uranium trioxide phases, the specialized version of PBE functional for solid materials, PBEsol [127], provided much better results [99, 100, 121]. The improved description of the structure of properties of anhydrous materials using this functional over the one obtained using PBE was observed in recent calculations [99, 100, 121] and later confirmed by other research groups [128, 129]. This justifies the great amount of effort made in developing density functionals specific for solid materials [130] and emphasizes the need of determining empirical dispersion parameters specific for these functionals. These functionals are implemented in CASTEP program [131], a module of the Materials Studio package [132], which was employed to model the structures of the materials considered. The pseudopotentials used for H, C, O, Si, and Ca atoms in the unit cells of these minerals were standard norm-conserving pseudopotentials [133] given in CASTEP code (00PBE-OP type). The norm-conserving relativistic pseudopotential for U atom was generated from first principles as shown in the previous works [101, 121]. While our uranium atom pseudopotential includes scalar relativistic effects, the corresponding pseudopotentials used for H, C, O, Si, and Ca atoms do not include them. This pseudopotential has been used extensively in the research of uranyl-containing materials [99–105, 120–123].

The atomic positions and cell parameters were optimized using the Broyden-Fletcher--Goldfarb-Shanno method [124, 134] with a convergence threshold on atomic forces of 0.01 eV/Å. The kinetic energy cut-off and K-point mesh [135] were chosen to ensure good convergence for computed structures and energies. The structures of the materials considered in this work were optimized in calculations with augmented complexity by increasing these parameters. The precise calculation parameters used to determine the final results may be found in the corresponding articles [99–105, 120–123]. The flow diagram associated to the theoretical treatment used to study a given crystalline material is shown in **Figure 1** [101]. The crystal structure is first optimized starting from an initial atomistic model of the corresponding unit cell (lattice parameters and atomic positions) employing trial values of the calculation parameters (kinetic energy cut-off and K-point mesh). The crystal structure is then updated and the initial values of the kinetic cutoff and K-point density (number of k-points divided by the k-point separation) are systematically increased. The geometry optimization is performed again until the variation of the energy is below a given threshold. The variation of the crystal unit cell is then analyzed, and the structure is reoptimized until this variation is small enough. Once the convergence in the computed energy and structure is met, the

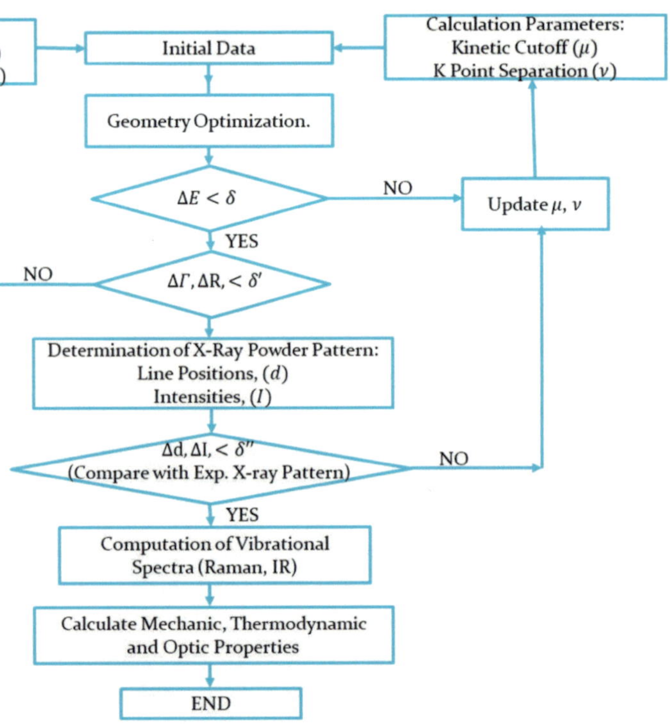

Figure 1. Flow diagram associated to the theoretical solid state treatment used to study the uranyl-containing materials considered in this work.

corresponding X-ray powder pattern is determined [136] and compared with the experimental one. Only if the comparison is satisfactory, the crystal structure is accepted in order to obtain the final vibrational, mechanic, thermodynamic, and optic properties of the material under study. Otherwise, the calculation parameters are made more stringent and the structural optimization starts again. The convergence of this procedure depends on the proximity of the initial input used to the final solution. If it does not converge or converges towards a structure yielding an X-ray powder pattern which does not agree with the experimental one, the procedure should be restarted from a different initial input (atomic positions and cell parameters).

2.2. Thermodynamic properties

The methods employed for the calculation of thermodynamic properties of these materials were described in the previous papers [99–105]. The phonon spectrum at the different points of Brillouin zone can be determined by density functional perturbation theory (DFPT) [137–139] as second-order derivatives of the total energy [137]. Phonon dispersion curves and density of states were calculated, and from them, several important thermodynamic quantities in the quasi-harmonic approximation, such as Gibbs free energy, enthalpy, entropy, and specific heat, were evaluated [140].

2.3. Enthalpy and Gibbs free energy of formation in terms of the elements

The thermodynamic functions of formation at the different temperatures were determined [102] from the calculated enthalpy and entropy functions of the material being considered, $[H_T\text{-}H_{298}]^{calc}$ and S_T^{calc}, the experimental value of its standard enthalpy of formation, $\Delta_f H^0$, and the experimental enthalpy and entropy functions of the elements forming part of the material. The enthalpy and entropy functions for the elements were taken from JANAF tables [11], and the corresponding functions for uranium atom were taken from Barin [12]. The equilibrium constants for the formation reactions were determined in terms of the corresponding calculated Gibbs free energies of formation using the well-known relationship [11], $\Delta_f G(T) = -RT\ Ln\ K_f$.

2.4. Enthalpies and free energies of reaction

The enthalpies and Gibbs free energies of a given reaction at the different temperatures, $\Delta_r H(T)$ and $\Delta_r G(T)$, were determined [103] from the Gibbs free energy of formation and entropy functions of the materials entering in the reaction, $\Delta_f G^i(T)$ and $S^i(T)$, $i = 1,\ldots,N_{mat}$. The specific values used of these properties for studtite, metastudtite, dehydrated schoepite, schoepite, metaschoepite, becquerelite, soddyite, rutherfordine, and gamma uranium trioxide were determined in our previous works [102, 104, 105]. The corresponding data for the remaining materials, which do not contain the uranyl ion, SiO_2(cr), H_2O(l), CO_2(g), O_2(g), and H_2(g), were taken from JANAF tables [11] and the data for H_2O_2(l) were taken from Barin [12]. The reaction equilibrium constants were determined in terms of the corresponding Gibbs free energies of reaction, $\Delta_r G(T) = -RT\ Ln\ K_r$.

3. Results and discussion

3.1. Thermodynamic properties

The computed isobaric heats and entropies at 298.15 K for all the uranyl-containing materials considered in this work [99, 100, 102, 104, 105] are given in **Table 1**. In this table, the values obtained for rutherfordine, gamma uranium trioxide, and metaschoepite [99, 100, 104] are compared with the corresponding experimental values. For the rest of the materials, there are not experimental values to compare with. As it can be seen, the computed values agree very well with their experimental counterparts. In fact, the differences between the computed and experimental values of these and other thermodynamic properties are frequently smaller than the difference among several different experimental values. From the analysis of the results obtained, the expected accuracy in the computed-specific heats and entropies of studtite, metastudtite, dehydrated schoepite, schoepite, and soddyite is better than 3–5%.

The calculated isobaric specific heat, entropy, and Gibbs free energy functions of rutherfordine, gamma uranium trioxide, and metaschoepite are displayed in **Figure 2**, where they are compared with the experimental functions of Hemingway [37], Cordfunke and Westrum [143], and Barin [12], respectively. For rutherfordine, the computed thermodynamic functions are compared with those of Hemingway [37] in the temperature range of 298–700 K, and as it can be appreciated, the calculated and experimental curves are nearly parallel. The computed value of C_p at 700 K, near the limit of thermal stability of rutherfordine [37], C_p = 153.3 J K^{-1} mol^{-1}, differs from the experimental value at this temperature, 147.6 J · K^{-1} mol^{-1}, by only 3.9%.

Material	Source	C_p	S
Rutherfordine	Calc. [99]	115.02	143.11
	Exp.	106.5 [37] (8.0%), 120.1 [141] (−5.1%)	142.70 [37] (0.3%), 139 [142] (3.0%)
γ-UO_3	Calc. [100]	77.36	92.96
	Exp.	81.67 [143] (−5.3%), 84.72 [37] (−8.7%)	96.11 [143] (−3.3%), 98.6 [144] (−5.7%)
Metaschoepite	Calc. [104]	142.01	166.24
	Exp.	154.40 [12] (−8.0%)	167.00 [12] (−0.5%)
Studtite	Calc. [102]	219.97, 211.17 [41]	232.12
Metastudtite	Calc. [102]	163.14, 155.81 [41]	179.27
Dehydrated schoepite	Calc. [102]	103.85	125.18
Schoepite	Calc. [104]	150.62	168.75
Soddyite	Calc. [102]	275.15	315.95
Becquerelite	Calc. [105]	148.40	172.34

All the values are given in units of J K^{-1} mol^{-1}. The percent difference of the theoretical and experimental results is given in parenthesis for the specific heats and entropies of rutherfordine, gamma uranium trioxide, and metaschoepite.

Table 1. Specific heats and entropies at 298.15 K for the uranyl-containing material studied in this work.

http://dx.doi.org/10.5772/intechopen.79558

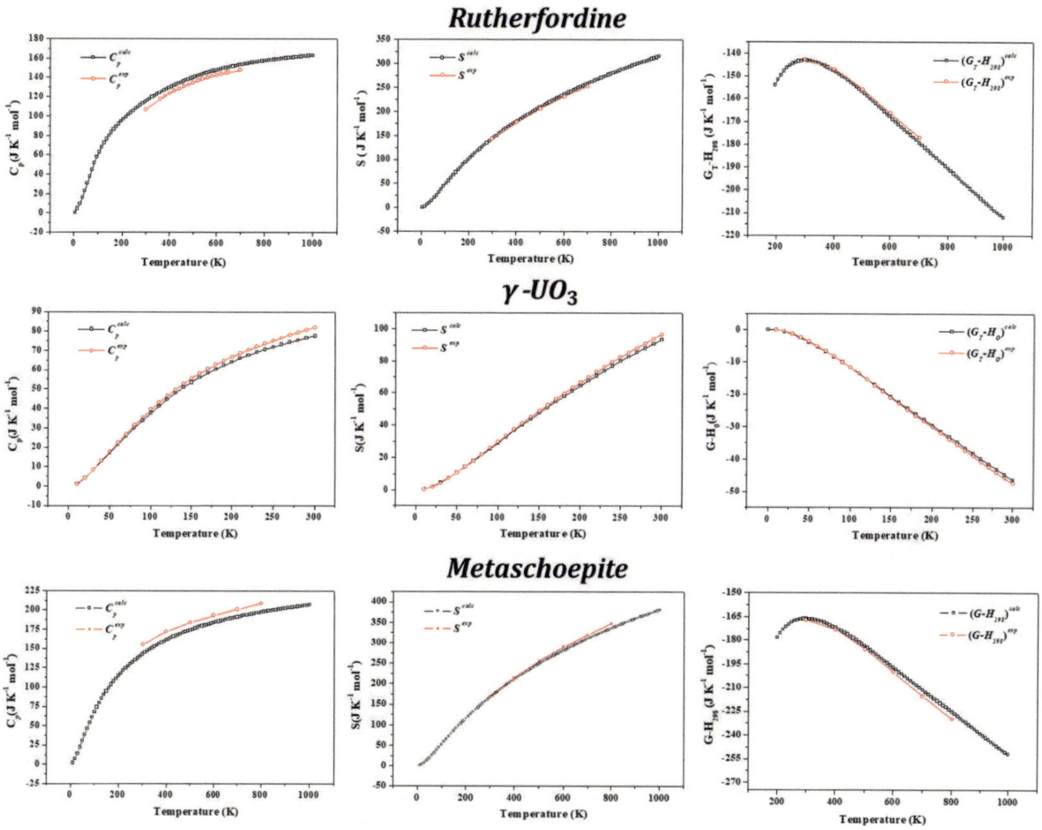

Figure 2. Calculated and experimental isobaric specific heat entropies and Gibbs free energies of rutherfordine, gamma uranium trioxide, and metaschoepite as a function of temperature. The experimental thermodynamic functions of rutherfordine, uranium trioxide, and metaschoepite are from Hemingway [37], Cordfunke and Westrum [143], and Barin [12], respectively.

Similarly, the differences of the computed entropy and Gibbs free energy with respect to Hemingway's experimental values at 700 K are only 2.3 and 1.3%, respectively. Our theoretical calculations allowed to obtain the values of the thermodynamic functions for the low and high temperature ranges 0–300 and 700–1000 K, which were unknown so far and, consequently, extended the range in which the thermodynamic functions were known to 0–1000 K.

The computed thermodynamic properties of uranium trioxide are also in excellent agreement with the experimental data of Cordfunke and Westrum [143] in the full range of temperatures considered 0–1000 K. The differences in the specific heat, entropy, and Gibbs free energy functions are 3.9, 1.8, and 0.1% at 100 K and 6.1, 3.6, and 3.5% at 1000 K. The comparison reveals that the low temperature calculated thermodynamic data are also very accurate. It must be emphasized that while the experimental isobaric heat capacity function of gamma uranium trioxide at 1000 K is above the asymptotic Dulong-Petit limit, our computed function satisfies properly the requirement of being below this limit [100].

Finally, the theoretical results for metaschoepite mineral phase agree very well with the experimental thermodynamic properties reported by Barin [12] even at temperatures of the order of 800 K, the percent differences of the calculated specific heat, entropy, and Gibbs energy with respect to the corresponding experimental values being 5.4%, 3.2%, and 2.0% at 800 K. The present theoretical data have permitted to discriminate between the experimental thermodynamic functions of metaschoepite reported up to date because the experimental functions reported by Tasker et al. [145] deviate from those of Barin [12] and from our theoretical results already at moderate temperatures [104].

The comparisons performed in the previous paragraph have shown that the variation of the computed thermodynamic functions with temperature is excellent. Hence, it may be expected that the theoretical functions obtained for studtite, metastudtite, dehydrated schoepite, soddyite, schoepite, and becquerelite [102, 104, 105] will also be reliable even at low and high temperatures. For rutherfordine and metaschoepite [99, 104], the calculated thermodynamic properties are recommended instead of the experimental functions because they cover the full temperature range going from 0 to 1000 K and they should provide a uniform accuracy at all temperatures. For gamma uranium trioxide, both sets of data are considered to be equally accurate, but the theoretical specific heat function satisfies properly the asymptotic conditions [100].

3.2. Enthalpies and free energies of formation in terms of the elements

The enthalpies and free energies of formation in terms of the elements of the considered mineral phases as a function or temperature were determined [102, 104, 105] from the calculated thermodynamic data, the experimental or estimated [104] standard enthalpy of formation, and the thermodynamic functions of the corresponding elements [11–12]. The calculated Gibbs free energies of formation of rutherfordine, gamma uranium trioxide, and metaschoepite [102, 104] are shown in **Figure 3** together with the corresponding experimental data [12, 37, 143].

As it may be observed in **Figure 3**, and as it occurred with the thermodynamic functions of the pure substances reported in Section 3.1, the calculated thermodynamic properties of formation agree with the experimental functions in an excellent manner. For these three materials, the differences of the calculated and experimental values are lower than 1% at ambient temperature and the differences remain very small at high temperatures. The differences become 1.6%, 1.0%, and 2.0% at 700, 900, and 800 K for rutherfordine, γ-UO_3, and metaschoepite, respectively [102, 104]. Since the theoretical solid state treatments used for studtite, metastudtite, dehydrated schoepite, soddyite, schoepite, and becquerelite were essentially the same as those used for these three materials, we expect a similar accuracy level for their calculated thermodynamic parameters of formation. An example of these calculated parameters for a material in which there are no experimental data to compare with is the case of the schoepite mineral phase, and the corresponding results are also shown in **Figure 3**. The combination of the results for schoepite and metaschoepite allowed to study the thermodynamics of the dehydration transformation of schoepite into metaschoepite [104].

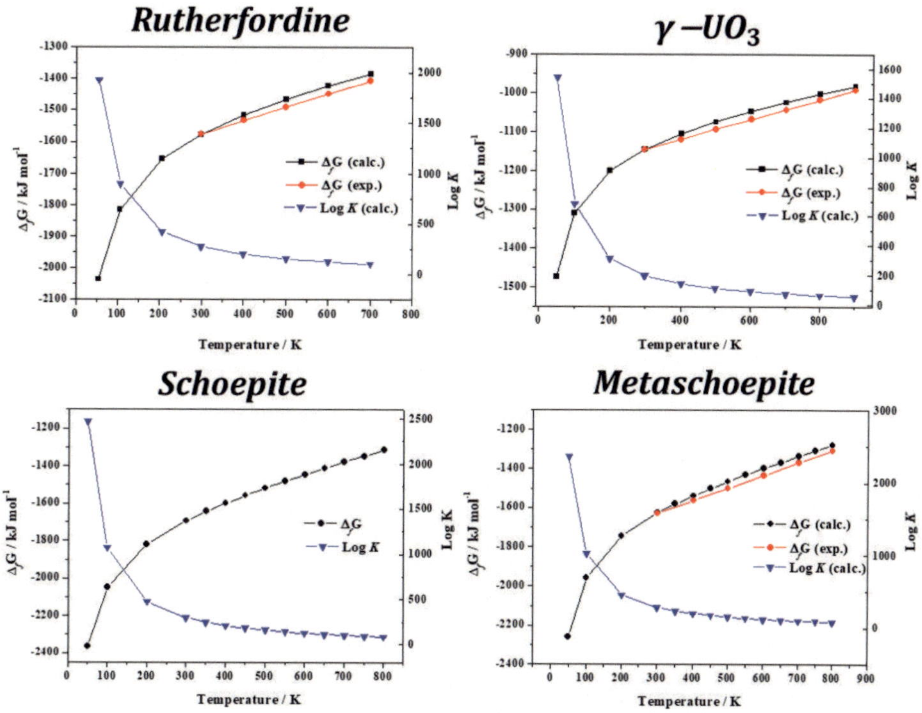

Figure 3. Calculated Gibbs free energies of formation of rutherfordine, gamma uranium trioxide, and metaschoepite in terms of the elements as a function of temperature. The experimental Gibbs free energies of formation of rutherfordine, uranium trioxide, and metaschoepite are from Hemingway [37], Cordfunke and Westrum [143], and Barin [12], respectively. For schoepite, there are no experimental data to compare with.

3.3. Enthalpies and free energies of reaction

3.3.1. Reactions of formation in terms of oxides

Let us first consider the following reactions:

$$\text{Studtite: } UO_3(cr) + 4\,H_2O(l) + 1/2O_2(g) \rightarrow (UO_2)O_2 \cdot 4H_2O(cr) \qquad (I)$$

$$\text{Metastudtite: } UO_3(cr) + 2\,H_2O(l) + 1/2O_2(g) \rightarrow (UO_2)O_2 \cdot 2H_2O(cr) \qquad (II)$$

$$\text{Dehydrated schoepite: } UO_3(cr) + H_2O(l) \rightarrow UO_2(OH)_2(cr) \qquad (III)$$

$$\text{Becquerelite: } UO_3(cr) + 1/6\,CaO(cr) + 11/6H_2O(l) \rightarrow 1/6\,Ca(UO_2)_6O_4(OH)_6 \cdot 8H_2O(cr) \qquad (IV)$$

$$\text{Schoepite: } UO_3(cr) + 9/4\,H_2O(l) \rightarrow 1/8\left[(UO_2)_8O_2(OH)_{12}\right] \cdot 12H_2O(cr) \qquad (V)$$

$$\text{Metaschoepite: } UO_3(cr) + 2\,H_2O(l) \rightarrow 1/8\left[(UO_2)_8O_2(OH)_{12}\right] \cdot 10\,H_2O(cr) \qquad (VI)$$

$$\text{Rutherfordine: } UO_3(cr) + CO_2(g) \rightarrow UO_2CO_3(cr) \qquad (VII)$$

$$\text{Soddyite : } 2UO_3(cr) + SiO_2(cr) + 2H_2O(l) \rightarrow (UO_2)_2(SiO_4) \cdot 2H_2O(cr) \qquad (VIII)$$

These important reactions represent the formation of the considered uranyl-containing materials in terms of the corresponding oxides. By combining the calculated thermodynamic properties of formation of these materials in terms of the elements [102, 104, 105] with those of the non-uranyl-bearing materials [11] present in reactions (I)–(VIII), we obtained the Gibbs free energies and associated reaction constants displayed in **Figure 4** [103–105].

Figure 4 shows that studtite and metastudtite are unstable with respect to the corresponding oxides at the full range of temperatures studied, 250–500 K, since the corresponding Gibbs free energies of reaction are positive everywhere. Therefore, they are metastable phases at normal conditions. The opposite is true for soddyite mineral phase, which is stable at all the temperatures. However, dehydrated schoepite, becquerelite, schoepite, metaschoepite, and rutherfordine mineral phases are stable at ambient temperature and become unstable at the temperatures of 462, 491, 383, 352, and 514 K, respectively, because the Gibbs free energy of reaction becomes positive at these temperatures. The observation of changes of stability for these phases at these relatively low temperatures was unexpected and highlights the great relevance of the availability of accurate temperature-dependent thermodynamic functions [102, 103].

3.3.2. Reactions of transformation of uranyl-containing materials into studtite in the presence of high hydrogen peroxide concentrations

We will now study the thermodynamic properties of the following set of reactions:

$$UO_2(OH)_2(cr) + 2\,H_2O_2(l) + H_2(g) \rightarrow (UO_2)O_2 \cdot 4H_2O(cr) \qquad (IX)$$

$$1/2\,(UO_2)_2(SiO_4) \cdot 2H_2O(cr) + 2\,H_2O_2(l) + H_2(g) \rightarrow (UO_2)O_2 \cdot 4H_2O(cr) + 1/2\,SiO_2(cr) \qquad (X)$$

$$UO_3(cr) + 3\,H_2O_2(l) + H_2(g) \rightarrow (UO_2)O_2 \cdot 4H_2O(cr) + 1/2\,O_2(g) \qquad (XI)$$

$$UO_2CO_3(cr) + 3\,H_2O_2(l) + H_2(g) \rightarrow (UO_2)O_2 \cdot 4H_2O(cr) + CO_2(g) + 1/2O_2(g) \qquad (XII)$$

$$(UO_2)O_2 \cdot 2H_2O(cr) + H_2O_2(l) + H_2(g) \rightarrow (UO_2)O_2 \cdot 4H_2O(cr) \qquad (XIII)$$

$$1/6\,Ca(UO_2)_6O_4(OH)_6 \cdot 8H_2O(cr) + 13/6\,H_2O_2(l) \rightarrow (UO_2)O_2 \cdot 4H_2O(cr) + 1/16\,CaO(cr) + 7/12\,O_2(g) \qquad (XIV)$$

$$1/8\,\left[(UO_2)_8O_2(OH)_{12}\right] \cdot 12H_2O(cr) + 7/4\,H_2O_2(l) \rightarrow (UO_2)O_2 \cdot 4H_2O(cr) + 3/8\,O_2(g) \qquad (XV)$$

$$1/8\,\left[(UO_2)_8O_2(OH)_{12}\right] \cdot 12H_2O(cr) + 1/8\,O_2(g) \rightarrow 1/8\,\left[(UO_2)_8O_2(OH)_{12}\right] \cdot 10\,H_2O(cr) + 1/4\,H_2O_2(l) \qquad (XVI)$$

Reactions (IX) to (XV) are the reactions of transformation of dehydrated schoepite, soddyite, uranium trioxide, rutherfordine, metastudtite, becquerelite, and schoepite into studtite in the presence of high hydrogen peroxide concentrations (and absence of water) [103–105]. Reaction (XVI) represents the conversion of schoepite into metaschoepite at these conditions. The computed Gibbs free energies of these reactions are shown in **Figure 5**. As it can be observed, all

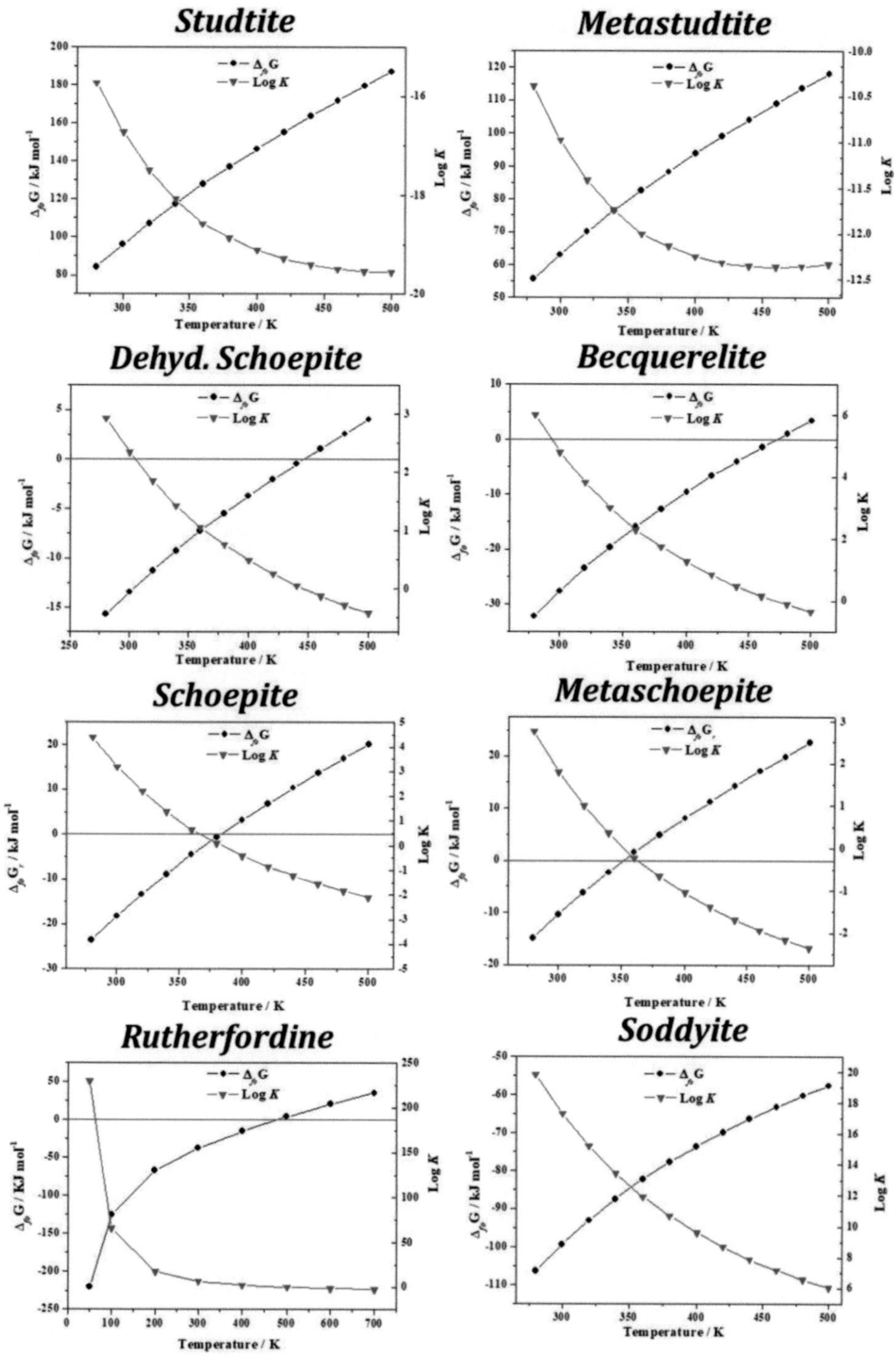

Figure 4. Calculated Gibbs free energies of formation of studtite, metastudtite, dehydrated schoepite, becquerelite, schoepite, metaschoepite, rutherfordine, and soddyite in terms of the corresponding oxides as a function of temperature [103–105].

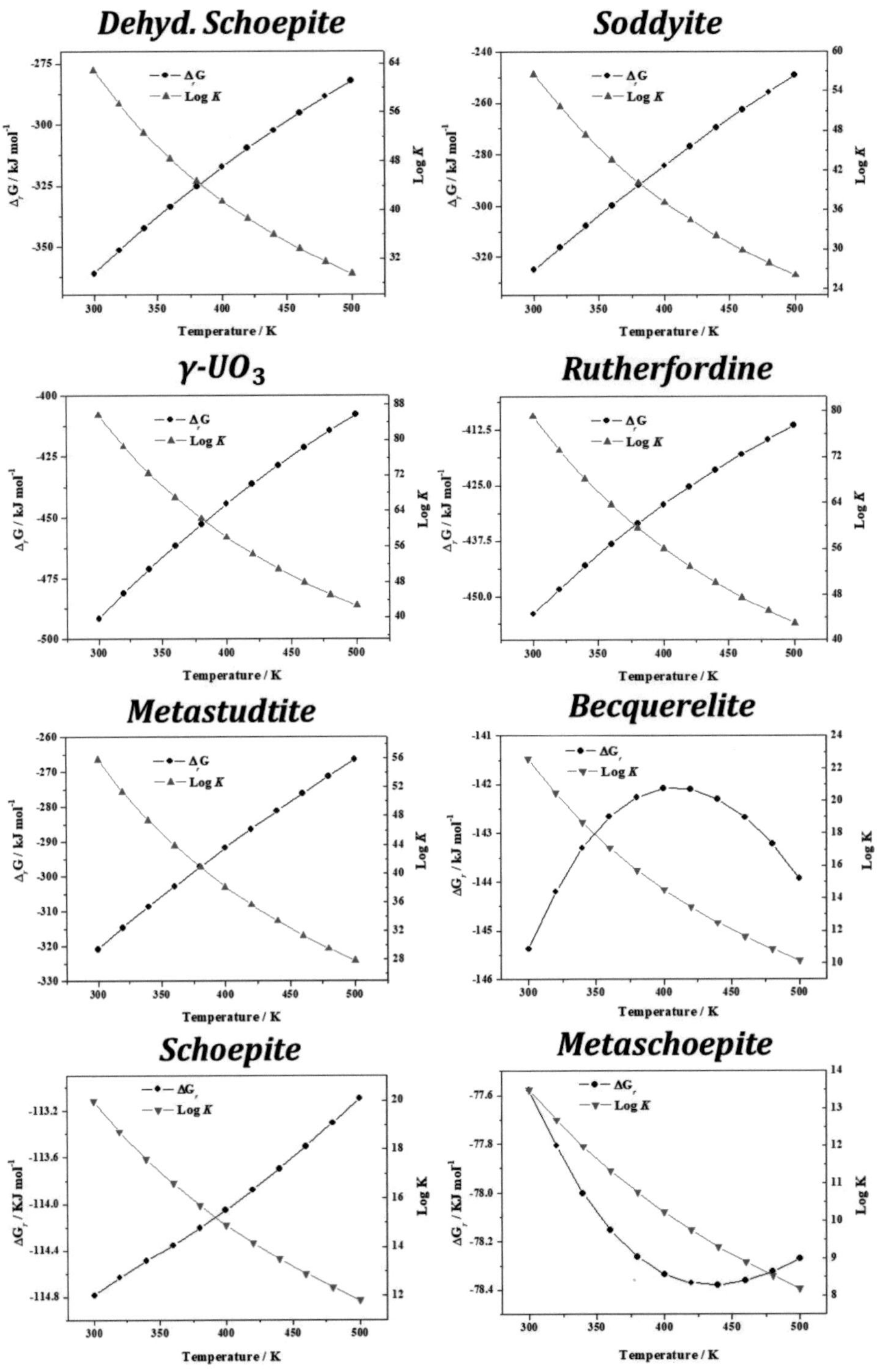

Figure 5. Calculated Gibbs free energies of the reactions of transformation of dehydrated schoepite, soddyite, gamma uranium trioxide, rutherfordine, metastudtite, becquerelite, and schoepite into studtite and of metaschoepite into schoepite in the presence of high hydrogen peroxide concentrations as a function of temperature [103–105].

these phases will transform spontaneously into studtite in the presence of high hydrogen peroxide concentrations, since the Gibbs free energy of all these reactions are negative for the full range of temperatures considered, 300–500 K. In the case of metaschoepite, the thermodynamics of reaction (XVI) shows that it will convert into schoepite mineral phase, but according to reaction (XV) the last phase also transforms spontaneously into studtite [104].

Forbes et al. [110] investigated experimentally the transformation of dehydrated schoepite and soddyite into studtite phase in solutions with large concentrations of hydrogen peroxide. They observed that, at ambient temperature, these materials transform into studtite following the reaction stoichiometry. The results of our calculations [103] for the conversion of dehydrated schoepite and soddyite under high hydrogen peroxide concentrations into studtite agree completely with this experimental study. However, our results extend this study because it shows that the same will happen not only at 298.15 K but also at temperature as high as 500 K. The study performed by Kubatko et al. [146] showed that becquerelite mineral phase also transforms into studtite within 8 hours under high hydrogen peroxide concentrations. The thermodynamics of the conversion of becquerelite into studtite under variable concentrations of hydrogen peroxide was studied by our group in a recent paper [105]. The results displayed in **Figure 4** [103–105] show that the same will happen for gamma uranium trioxide, rutherfordine, metastudtite, schoepite, and metaschoepite phases. In fact, because the stability of studtite under these conditions is very high, it is likely that the same will happen for most of the other secondary phases of SNF, as it was suggested in 2017 [101].

Our study of the thermodynamics of these reactions also permits to comprehend why uranyl peroxide hydrates were the unique phases found in a 2-year corrosion experiment of SNF in deionized water [109]. These phases should be the unique phases found not only in deionized water but also in water containing silicate ions, since studtite is much more stable than soddyite and probably more stable than most other uranyl silicate phases under high hydrogen peroxide concentrations.

3.4. Thermodynamic stability

From the thermodynamic data reported in our previous papers [103–105], the order of thermodynamic stability of the uranyl-containing materials considered in this work was evaluated as a function of temperature under three different conditions: (A) under high concentrations of hydrogen peroxide; (B) in the presence of water and hydrogen peroxide; and (C) in the absence of hydrogen peroxide. The stability of these phases at these conditions in the range of temperatures from 300 to 500 K is displayed in **Figures 5A** and **6B** and **C**. In these three figures, the relative stabilities are given with respect to studtite, metastudtite, and gamma uranium trioxide, respectively.

Figure 6 provides a very clear idea of the temporal evolution of the paragenetic sequence of secondary phases appearing as a result of the corrosion of SNF under final DGR conditions. Uranyl peroxide phase studtite will appear as the prominent phase at the earlier stages of this paragenetic sequence (see **Figure 6A**) due to the presence of high hydrogen peroxide concentrations caused by the radiolysis of most of the water reaching the surface of SNF. If the hydrogen peroxide concentration diminishes with time, as expected from the decrease of the

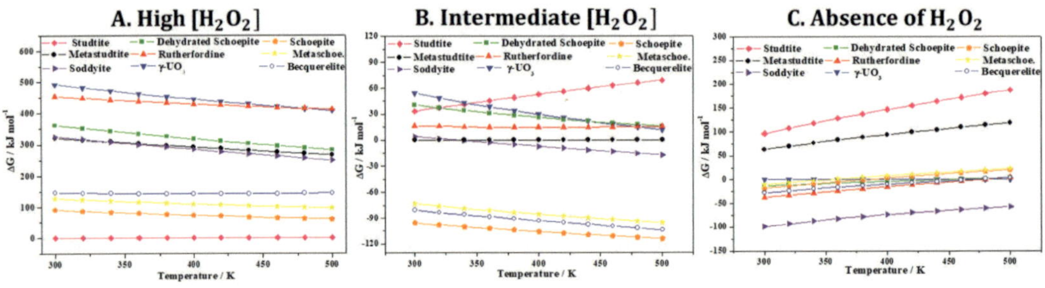

Figure 6. Relative thermodynamic stability of a selected set of secondary phases of SNF: (A) under high hydrogen peroxide concentrations; (B) under the presence of water and hydrogen peroxide; and (C) under the absence of hydrogen peroxide.

intensity of radiation fields over time in a DGR [147], the studtite stability will decrease and the formation of other secondary phases will occur. In the presence of water and hydrogen peroxide (see **Figure 6B**), the uranyl oxyhydroxide phases (schoepite, metaschoepite, and becquerelite) appear to be the most stable ones. Finally, in the absence of hydrogen peroxide, soddyite is the most stable phase and rutherfordine is also more stable than becquerelite for temperatures lower than 492 K (**Figure 6C**). Thus, at hydrogen peroxide free conditions, uranyl silicates and carbonates must be the most prominent phases of the SNF.

A full evaluation and understanding of the number and relative amount of the secondary phases of spent nuclear fuel present at the conditions of a final geological disposal over time require the realization of complete thermodynamic calculations employing thermochemical data for a significant number of materials, including the most important secondary phases, amorphous phases, and aqueous species, at a wide range of temperature and pressure conditions [103]. The determination of these thermodynamic data, the evaluation of their temperature and pressure dependence, and the realization of the corresponding thermodynamic computations are one of the main objectives of our current research.

3.5. Solubility constants

The important solubility reactions of schoepite, metaschoepite, rutherfordine, and becquerelite may be written, respectively, as follows:

$$1/8\ \left[(UO_2)_8O_2(OH)_{12}\right]\cdot 12H_2O(cr) + 2\,H^+(aq) \rightarrow {UO_2}^{2+}(aq) + 13/4\ H_2O(l) \qquad \text{(XVII)}$$

$$1/8\ \left[(UO_2)_8O_2(OH)_{12}\right]\cdot 10H_2O(cr) + 2\,H^+(aq) \rightarrow {UO_2}^{2+}(aq) + 3\ H_2O(l) \qquad \text{(XVIII)}$$

$$Ca(UO_2)_6O_4(OH)_6\cdot 8H_2O(cr) + 14\,H+(aq) \rightarrow Ca^{2+}(aq) + 6\,{UO_2}^{2+}(aq) + 16\ H_2O(l) \qquad \text{(XIX)}$$

$$UO_2CO_3(cr) \rightarrow {UO_2}^{2+}(aq) + {CO_3}^{2-}(aq) \qquad \text{(XX)}$$

Using the computed values of the Gibbs free energies of formation of schoepite, metaschoepite, becquerelite, and rutherfordine and the Gibbs free energies of formation of aqueous ions, ${UO_2}^{2+}(aq)$,

Material	$\Delta_{sp}G$ (calc.)	Log K_{sp} (calc.)	Log K_{sp} (exp.)
Schoepite	−26.11	4.57	—
Metaschoepite	−34.14	5.98	5.6 ± 0.2 [149], 5.52 ± 0.04 [150], 6.23 ± 0.14 [151], 5.9 ± 0.1 [152], 5.14 ± 0.05 [153], 5.72 ± 0.19 [154], 5.79 ± 0.19 [155]
Becquerelite	−287.55	50.38	40.5 ± 1.4 [149], 41.2 ± 0.52 [156], 43.2 [157], 29 ± 1 [158], 41.89 ± 0.52 [159], 43.70 ± 0.47 [159]
Rutherfordine	96.83	−16.96	−14.91 ± 0.10 [153], −13.89 ± 0.11 [154], −13.29 ± 0.01 [155]

The values of $\Delta_{sp}G$ Δ_rG are in units of $kJ{\cdot}mol^{-1}$.

Table 2. Calculated and experimental Gibbs free energies ($\Delta_{sp}G$) and associated reaction constants (Log K_{sp}) of the solubility reactions of schoepite, metaschoepite, becquerelite, and rutherfordine.

UO_2^{2+} (aq), CO_3^{2-}(aq), Ca^{2+}(aq), and H^+(aq), and liquid water at 298.15 K [148], one obtains the Gibbs free energies and associated reaction constants of solubility given in **Table 2**.

The calculated solubility products, $LogK_{sp}^{calc}$ of metaschoepite, becquerelite, and rutherfordine 5.98, 50.38, and −16.96 respectively, are in very good agreement with the most recent experimental values ($LogK_{sp}^{exp}$ = 5.6 ± 0.2 [149], 40.5 ± 1.4 [149], −14.91 ± 0.10 [153]). Since there solubility constant of schoepite has not been determined experimentally, its value was predicted [104]. Schoepite is shown to be more insoluble than metaschoepite.

4. Conclusions

It has been demonstrated [99–105, 120–123] that Periodic Density Functional Theory methods are an extremely powerful tool in the research of uranium-containing compounds. The use of the new relativistic norm-conserving pseudopotential [101, 121] permitted the computation of the structural properties, X-ray powder patterns, vibrational Raman spectra, and mechanical and thermodynamic properties of these materials. These methods are free of the problems involved in the experimental methods associated to the radiotoxicity of these compounds.

The first principles methodology allowed the safe, accurate, and cheap study of secondary phases of SNF definitive geological disposal conditions. The theoretical methods may be used, in conjunction with experimental techniques, as an interpretative tool of the experimental data or as a predictive tool to determine the structural, vibrational, mechanic, and thermodynamic properties of these substances. One of the most successful applications of this methodology has been achieved when studying their fundamental thermodynamic properties [99–105].

The development of empirical dispersion corrections [126] and the development of density functionals specific for solid materials [130] have improved extraordinarily the reliability of the calculated thermodynamic functions and their temperature dependence. The results were shown to be accurate at very low and high temperatures [99–105]. The description of the temperature dependence of these functions is very difficult from the experimental point of view. The theoretical approach has permitted in some cases, as those of the rutherfordine [99] and metaschoepite [104] mineral phases, to extend the range of temperatures in which the thermodynamic properties were known and to determine the variation with temperature of these properties for a large series of important phases in which it was completely unknown: studtite, metastudtite, dehydrated schoepite, becquerelite, schoepite, and soddyite. Furthermore, the calculated thermodynamic functions satisfy properly the Dulong-Petit asymptotic constraints.

The comparison of the computed heat capacities and entropies with experimental data was very satisfactory in those cases in which there ware experimental data to compare with. The calculated Gibbs free energies of formation of rutherfordine, γ-UO_3, and metaschoepite [102, 104] were in good agreement with experiment at ambient temperature, and the differences with the corresponding experimental values were only 1.6%, 1.0%, and 2.0% at 700, 900, and 800 K, respectively. Because the theoretical treatments used for studtite, metastudtite, dehydrated schoepite, soddyite, schoepite, and becquerelite were essentially the same as those used for these three materials, we expect a similar accuracy level for their calculated thermodynamic parameters of formation [102, 104, 105].

As an application of the calculated thermodynamic properties of the considered uranyl materials, the Gibbs free energies and associated reaction constants of a large number of reactions involving these materials were determined. The results provided a deep and clear understanding of the temporal evolution of the paragenetic sequence of secondary phases appearing at the surface of SNF as a result of its corrosion under final DGR conditions [103–105]. Additional work is now in progress to determine the thermodynamic properties of a significant number of additional phases. The use of these thermodynamic parameters in detailed multi-component thermodynamic computations should be pursued in a near future.

Acknowledgements

Supercomputer time by the CETA-CIEMAT, CTI-CSIC, and CESGA centers is acknowledged. This work has been carried out in the context of a CSIC–CIEMAT collaboration agreement: "Caracterización experimental y teórica de fases secundarias y óxidos de uranio formados en condiciones de almacenamiento de combustible nuclear." I want to thank Dr. Ana María Fernández, Dr. Vicente Timon, Dr. Laura J. Bonales, Dr. Joaquín Cobos, and Dr. Rafael Escribano for their continuous help and advice during the realization of these studies.

Conflict of interest

The author declares that there is no conflict of interest.

http://dx.doi.org/10.5772/intechopen.79558

Dedication

To Beatriz and Ana María, on the occasion of their 12th and 25th birthdays.

Permissions

Rutherfordine. Part of **Figure 2** and data contained in **Table 1** adapted from Ref. [99] with permission from American Chemical Society, Copyright (2017). Part of **Figure 3** adapted from Ref. [102] with permission from American Chemical Society, Copyright (2018). Parts of **Figures 4–6** adapted from Ref. [103] with permission from American Chemical Society, Copyright (2018).

Gamma uranium trioxide. Part of **Figure 2** and data contained in **Table 1** adapted from Ref. [100] with permission from American Chemical Society, Copyright (2017). Part of **Figure 3** adapted from Ref. [102] with permission from American Chemical Society, Copyright (2018). Parts of **Figures 4–6** adapted from Ref. [103] with permission from American Chemical Society, Copyright (2018).

Dehydrated schoepite, studtite, metastudtite, and soddyite. Data contained in **Table 1** adapted from Ref. [102] with permission from American Chemical Society, Copyright (2018). Parts of **Figures 4–6** adapted from Ref. [103] with permission from American Chemical Society, Copyright (2018).

Schoepite and metaschoepite. Parts of **Figures 2–6** and data contained in **Tables 1 and 2** adapted from Ref. [104] with permission from American Chemical Society.

Becquerelite. Part of **Figures 4–6** and data contained in **Tables 1** and **2** reproduced from Ref. [105] with permission from the Royal Society of Chemistry.

Author details

Francisco Colmenero Ruiz

Address all correspondence to: francisco.colmenero@iem.cfmac.csic.es

Instituto de Estructura de la Materia – Consejo Superior de Investigaciones Científicas (IEM-CSIC), Madrid, Spain

References

[1] Guggenheim EA. Thermodynamics: An Advanced Treatment for Chemists and Physicists. Amsterdam: North-Holland; 1967. ISBN: 0444869514

[2] Gorman-Lewis D, Burns PC, Fein JB. Review of uranyl mineral solubility measurements. Journal of Chemical Thermodynamics. 2008;**40**:335-352. DOI: 10.1016/j.jct.2007.12.004

[3] Shvareva TY, Fein JB, Navrotsky A. Thermodynamic properties of uranyl minerals: Constraints from calorimetry and solubility measurements. Industrial and Engineering Chemical Research. 2012;**51**:607-613. DOI: 10.1021/ie2002582

[4] Gasparik T. An internally consistent thermodynamic model for the system CaO–MgO–Al_2O_3–SiO_2 derived primarily from phase equilibrium data. The Journal of Geology. 2000;**108**:103-119. DOI: 10.1086/314389

[5] Bruno J, Bosbach D, Kulik D, Navrotsky A. Chemical thermodynamics of solid solutions of interest in radioactive waste management. In: Mompean FJ, Illemassene M, editors. Issy-Les-Moulineaux, France: OECD Nuclear Energy Agency, Data Bank; 2007. DOI: 10.1787/9789264033191-en

[6] Chopelas A. Thermal properties of forsterite at mantle pressures derived from vibrational spectroscopy. Physics and Chemistry of Minerals. 1990;**17**:149-156. DOI: 10.1007/BF00199666

[7] Chopelas A. Thermal expansion, heat capacity, and entropy of MgO at mantle pressures. Physics and Chemistry of Minerals. 1990;**17**:142-148. DOI: 10.1007/BF00199665

[8] Hofmeister AM, Chopelas A. Thermodynamic properties of pyrope and grossular from vibrational spectroscopy. American Mineralogist. 1991;**76**:880-891. ISSN: 0003-004X

[9] Helgeson HC, Delany JM, Nesbit HW, Bird DK. Summary and critique of the thermodynamic properties of rock-forming minerals. American Journal of Science. 1978;**278A**:1-229. ISSN: 0002-9599

[10] Wagman DD, Evans WH, Parker VB, Schumm RH, Halow I, et al. The NBS tables of chemical thermodynamic properties. Selected values for inorganic and C1 and C2 organic substances in SI units. Journal of Physical Chemistry Reference Data. 1982;**11** (Suppl. 2):1-392. ISBN: 0-88318-417-6

[11] Chase MW, Davies CA, Downey JR, Frurip DJ, McDonald RA, Syverud AN. JANAF Thermochemical Tables. Third Edition. Journal of Physical Chemistry Reference Data. 1985;**14**(Suppl. 1):1-1856. ISBN: 0-88318-473-7

[12] Barin I. Thermochemical Data of Pure Substances. Third ed. Weinheim: VCH; 1995. DOI: 10.1002/9783527619825

[13] Barin I, Schmidt W, Schilling J. EQUITHERM Database and Software Package for Chemical Equilibrium Calculations on Personal Computers. Munster: VCH Scientific Software; 1993. ISBN: 352726373X

[14] Berman RG. Internally-consistent thermodynamic data for minerals in the system Na_2O-K_2O-CaO-MgO-FeO-Fe_2O_3-Al_2O_3-SiO_2-TiO_2-H_2O-CO_2. Journal of Petrology. 1988;**29**: 445-522. DOI: 10.1093/petrology/29.2.445

[15] Holland TJB, Powell R. An enlarged and updated internally consistent thermodynamic dataset with uncertainties and correlations: The system K_2O-Na_2O-CaO-MgO-MnO-FeO-Fe_2O_3-Al_2O_3-TiO_2-SiO_2-C-H_2-O_2. Journal of Metamorphic Geology. 1990;**8**:89-124. DOI: 10.1111/j.1525-1314.1990.tb00458.x

[16] Holland TJB, Powell R. An internally consistent thermodynamic data set for phases of petrological interest. Journal of Metamorphic Geology. 1998;**16**:309-343. DOI: 10.1111/j.1525-1314.1998.00140.x

[17] Holland TJB, Powell R. An improved and extended internally consistent thermodynamic dataset for phases of petrological interest, involving a new equation of state for solids. Journal of Metamorphic Geology. 2011;**29**:333-383. DOI: 10.1111/j.1525-1314.2010.00923.x

[18] Bale CW, Belisle E, Chartrand P, Decterov SA, Eriksson G, Gheribi AE, et al. FactSage thermochemical software and databases, 2010–2016. Calphad. 2016;**54**:35-53. DOI: 10.1016/j.calphad.2016.05.002

[19] Lukas HL, Henig ET, Zimmermann B. Optimization of phase diagrams by a least squares method using simultaneously different types of data. Calphad. 1977;**1**:225-236. DOI: 10.1016/0364-5916(77)90002-5

[20] Van de Walle A, Ceder G. Automating first-principles phase diagram calculations. Journal of Phase Equilibria. 2002;**23**:348-359. DOI: 10.1361/10549710277

[21] Fabrichnaya O, Saxena SK, Richet P, Westrum EF. Thermodynamic Data, Models, and Phase Diagrams in Multicomponent Oxide Systems. Berlin: Springer-Verlag; 2004. ISBN: 3540140182

[22] Eriksson G. Thermodynamic studies of high-temperature equilibria. XII. SOLGASMIX, a computer-program for calculation of equilibrium compositions in multiphase systems. Chemica Scripta. 1975;**8**:100-103. ISSN: 0004-2056

[23] George B, Brown LP, Farmer CH, Buthod P, Manning FS. Computation of multicomponent, multiphase equilibrium. Industrial and Engineering Chemical Process Design and Development. 1976;**15**:372-377. DOI: 10.1021/i260059a003

[24] Smith WR, Missen RW. Chemical Reaction Equilibrium Analysis: Theory and Algorithms. New York: Wiley-Interscience; 1982. ISBN: 0471093475

[25] Harvie CE, Greenberg JP, Weare JH. A chemical equilibrium algorithm for highly nonideal multiphase systems: Free energy minimization. Geochimica et Cosmochimica Acta. 1987;**51**:1045-1057. DOI: 10.1016/0016-7037(87)90199-2

[26] Piro MHA, Simunovic S. Performance enhancing algorithms for computing thermodynamic equilibria. Calphad. 2012;**39**:104-110. DOI: 10.1016/j.calphad.2012.09.005;

[27] Piro MHA, Simunovic S. Global optimization algorithms to compute thermodynamic equilibria in large complex systems with performance considerations. Computational Material Science. 2016;**118**:87-96. DOI: 10.1016/j.commatsci.2016.02.043

[28] Piro MHA. Updating the estimated assemblage of stable phases in a Gibbs energy minimizer. Calphad. 2017;**58**:115-121. DOI: 10.1016/j.calphad.2017.06.002

[29] Sundman B, Jansson B, Andersson JO. The thermo-calc databank system. Calphad. 1985; **9**:153-190. DOI: 10.1016/0364-5916(85)90021-5

[30] Eriksson G, Hack K. ChemSage—A computer program for the calculation of complex chemical equilibria. Metal Transactions B. 1990:**21**;1013-1023. DOI: 10.1007/BF02670272.

[31] Piro MHA, Simunovic S, Besmann TM, Lewis BJ, Thompson WT. The thermochemistry library Thermochimica. Computational Materials Science. 2013;**67**:266-272. DOI: 10.1016/ j.commatsci.2012. 09.011

[32] Miron GD, Kulik DA, Dmytrieva SV, Wagner T. GEMSFITS: Code package for optimization of geochemical model parameters and inverse modeling. Applied Geochemistry. 2015;**55**:28-45. DOI: 10.1016/j.apgeochem.2014.10.013

[33] Saxena SK, Eriksson G. Theoretical computation of mineral assemblages in pyrolite and lherzolite. Journal of Petrology. 1983;**24**:538-555. DOI: 10.1093/petrology/24.4.538

[34] Saxena SK, Eriksson G. High temperature phase equilibria in a solar-composition gas. Geochimica et Cosmochimica Acta. 1983;**47**:1865-1874. DOI: 10.1016/0016-7037(83)90203-X

[35] Saxena SK, Eriksson G. Low- to medium-temperature phase equilibria in a gas of solar composition. Earth and Planetary Science Letters. 1983;**65**:7-16

[36] Saxena SK, Eriksson G. Chemistry of formation of terrestrial planets. Advances in Physical Geochemistry. 1986;**6**:30-101. DOI: 10.1016/0012-821X(83)90185-1

[37] Hemingway BS. Thermodynamic Properties of Selected Uranium Compounds and Aqueous Species at 298.15 K and 1 Bar and at Higher Temperatures—Preliminary Models for the Origin of Coffinite Deposits. USGS Open-File Report 82–619; 1982. Available from: https://pubs.usgs.gov/of/1982/0619/report.pdf [Accessed: 2018-05-15]

[38] Langmuir D. Aqueous Environmental Geochemistry. New Jersey: Prentice-Hall; 1997. pp. 486-557. ISBN: 0023674121

[39] Casas I, Bruno J, Cera E, Finch RJ, Ewing RC. Kinetic and Thermodynamic Studies of Uranium Minerals Assessment of the Long-Term Evolution of Spent Nuclear Fuel. SKB Technical Report 94–16. Stockholm: Swedish Nuclear Fuel and Waste Management Co.; 1994. Available from: http://www.skb.com/publication/10579/TR94-16webb.pdf [Accessed: 2018-05-15]

[40] Navrotsky A, Shvareva TY, Guo X, Rock PA. Chapter 5: Thermodynamics of Uranium Minerals and Related Materials. In: Burns PC, Sigmon GE, editors. Uranium – Cradle to Grave. Mineralogical Association of Canada Short Course 43. Mineralogical Association of Canada: Québec; 2013. ISBN: 0921294530

[41] Sassani DC, Jové-Colón CF, Weck PF, Jerden JL, Frey KE, Cruse T, Ebert WL, Buck EC, Wittman RS. Used Fuel Degradation: Experimental and Modeling Report. Fuel Cycle

Research and Development Report FCRD-UFD-2013-000404. Sandia National Laboratories: Albuquerque, New Mexico; 2013. Available from: https://www.energy.gov/ne/downloads/used-fuel-degradation-experimental-and-modeling-report [Accessed: 2018-05-15]

[42] Frondel C. Mineral composition of gummite. American Mineralogist. 1956;**41**:539-568. ISSN: 0003-004X

[43] Frondel C. Systematic mineralogy of uranium and thorium. U.S. Geological Survey Bulletin. 1958;**1064**:1-400. Available from: http://pubs.er.usgs.gov/publication/b1064ER [Accessed: 2018-05-15]

[44] Finch RJ, Ewing RC. The corrosion of uraninite under oxidizing conditions. Journal of Nuclear Materials. 1992;**190**:133-156. DOI: 10.1016/0022-3115(92)90083-W

[45] Pearcy EC, Prikryl JD, Murphy WM, Leslie BW. Alteration of uraninite from the Nopal I deposit, Peña Blanca District, Chihuahua, Mexico, compared to degradation of spent nuclear fuel in the proposed US high-level nuclear waste repository at Yucca Mountain, Nevada. Applied Geochemistry. 1994;**9**:713-732. DOI: 10.1016/0883-2927(94)90030-2

[46] Wronkiewicz DJ, Bates JK, Gerding TJ, Veleckis E, Tani BS. Uranium release and secondary phase formation during unsaturated testing of UO_2 at 90°C. Journal of Nuclear Materials. 1992;**190**:107-127. DOI: 10.1016/0022-3115(92)90081-U

[47] Wronkiewicz DJ, Bates JK, Wolf SF, Buck EC. Ten-year results from unsaturated drip tests with UO2 at 90°C: Implications for the corrosion of spent nuclear fuel. Journal of Nuclear Materials. 1996;**238**:78-95. DOI: 10.1016/S0022-3115(96)00383-2

[48] Finch RJ, Murakami T. Systematics and paragenesis of uranium minerals. In: Ribbe PH, editor. URANIUM: Mineralogy, Geochemistry and the Environment; Reviews in Mineralogy, Vol. 38. New York: Mineralogical Society of America; 1999. pp. 91-180. ISBN: 0-939950-50-2

[49] Clark DL, Hobart DE, Neu MP. Actinide carbonate complexes and their importance in actinide environmental chemistry. Chemical Reviews. 1995;**95**:25-48. DOI: 10.1021/cr00033a002

[50] Sattonnay G, Ardois C, Corbel C, Lucchini JF, Barthe MF, Garrido F, Gosset D. Alpha-radiolysis effects on UO_2 alteration in water. Journal of Nuclear Materials. 2001;**288**:11-19. DOI: 10.1016/S0022-3115(00)00714-5

[51] Kubatko KA, Helean KB, Navrotsky A, Burns PC. Stability of peroxide-containing uranyl minerals. Science. 2003;**302**:1191-1193. DOI: 10.1126/science.1090259

[52] Kubatko KA. Crystallography, hierarchy of crystal structures, and chemical thermodynamics of selected uranyl phases [PhD thesis]. Illinois: Graduate School of the University of Notre Dame; 2005. Available from: https://www.researchgate.net/publication/277794995_Crystallography_Hierarchy_of_Crystal_Structures_and_Chemical_Thermodynamics_of_Select_Uranyl_Phases [Accessed: 2018-05-15]

[53] Guo X, Ushakova SV, Labs S, Curtius H, Bosbach D, Navrotsky A. Energetics of metastudtite and implications for nuclear waste alteration. Proceedings of the National Academy of Sciences of the United States of America. 2014;**111**:17737-17742. DOI: 10.1073/pnas.1421144111

[54] Armstrong CR, Nyman M, Shvareva TY, Sigmon GE, Burns PC, Navrotsky A. Uranyl peroxide enhanced nuclear fuel corrosion in seawater. Proceedings of the National Academy of Sciences of the United States of America. 2012;**109**:1874-1877. DOI: 10.1073/pnas.1119758109

[55] Burns PC, Ewing RC, Miller ML. Incorporation mechanisms of actinide elements in to the structures of U6+ phases formed during the oxidation of spent nuclear fuel. Journal of Nuclear Materials. 1997;**245**:1-9. DOI: 10.1016/S0022-3115(97)00006-8

[56] Burns PC. The crystal chemistry of uranium. In: Ribbe PH, editor. URANIUM: Mineralogy, Geochemistry and the Environment; Reviews in Mineralogy, Vol. 38. New York: Mineralogical Society of America; 1999. pp. 23-90. ISBN: 0-939950-50-2

[57] Burns PC. U6+ minerals and inorganic compounds: Insights into an expanded structural hierarchy of crystal structures. The Canadian Mineralogist. 2005;**43**:1839-1894. DOI: 10.2113/gscanmin.43.6.1839

[58] Burns PC. The structure of Boltwoodite and implications of solid solution towards sodium Boltwoodite. The Canadian Mineralogist. 1998;**36**:1069-1075. ISSN: 0008-4476

[59] Burns PC. Cs Boltwoodite obtained by ion exchange from single crystals: Implications for radionuclide release in a nuclear repository. Journal of Nuclear Materials. 1999;**265**:218-223. DOI: 10.1016/S0022-3115(98)00646-1

[60] Burns PC, Li Y. The structures of becquerelite and Sr-exchanged becquerelite. American Mineralogist. 2002;**87**:550-557. DOI: 10.2138/am-2002-0418

[61] Klingensmith AL, Burns PC. Neptunium substitution in synthetic Uranophane and Soddyite. American Mineralogist. 2007;**92**:1946-1951. DOI: 10.2138/am.2007.2542

[62] Burns PC, Deely KM, Skanthakumar S. Neptunium incorporation into uranyl compounds that form as alteration products of spent nuclear fuel: Implications for geologic repository performance. Radiochimica Acta. 2004;**92**:151-159. DOI: 10.1524/ract.92.3.151.30491

[63] Burns PC, Miller ML, Ewing RC. U6+ minerals and inorganic phases: A comparison and hierarchy of crystal structures. The Canadian Mineralogist. 1996;**34**:845-880. ISSN: 0008-4476

[64] Burns PC, Ewing RC, Hawthorne FC. The crystal chemistry of hexavalent uranium: Polyhedron geometries, bond-valence parameters, and polymerization of polyhedra. The Canadian Mineralogist. 1997;**35**:1551-1570. ISSN: 0008-4476

[65] Douglas M, Clark SB, Utsunomiya S, Ewing RC. Cesium and strontium incorporation into uranophane, $Ca[(UO_2)(SiO_3OH)]_2 \cdot 5H_2O$. Journal of Nuclear Science and Technology Supplement. 2002;**3**:504-507. DOI: 10.1080/00223131.2002.10875517

[66] Douglas M, Clark SB, Friese JI, Arey BW, Buck EC, Hanson BD, Utsunomiya S, Ewing RC. Microscale characterization of uranium(VI) silicate solids and associated neptunium(V). Radiochimica Acta. 2005;**93**:265-272. DOI: 10.1524/ract.93.5.265.64281

[67] Murphy WM, Grambow B. Thermodynamic interpretation of neptunium coprecipitation in uranophane for application to the Yucca Mountain repository. Radiochimica Acta. 2008;**96**:563-567. DOI: 10.1524/ract.2008.1537

[68] Shuller LC, Ewing RC, Becker U. Quantum-mechanical evaluation of Np-incorporation into studtite. American Mineralogist. 2010;**95**:1151-1160. ISSN: 0003-004X

[69] Shuller LC, Ewing RC, Becker U. Np-incorporation into uranyl phases: A quantum–mechanical evaluation. Journal of Nuclear Materials. 2013;**434**:440-450. DOI: 10.1016/j.jnucmat.2011.04.016

[70] Shuller LC, Bender WM, Walker SM, Becker U. Quantum-mechanical methods for quantifying incorporation of contaminants in proximal minerals. Minerals. 2014;**4**:690-715. DOI: 10.3390/min4030690

[71] Gorman-Lewis D, Mazeina L, Fein JB, Szymanowski JES, Burns PC, Navrotsky A. Thermodynamic properties of soddyite from solubility and calorimetry measurements. Journal of Chemical Thermodynamics. 2007;**39**:568-575. DOI: 10.1016/j.jct.2006.09.005

[72] Grenthe I, Fuger J, Konings RJM, Lemire RJ, Muller AB, Nguyen-Trung C, Wanner H. Chemical Thermodynamics of Uranium. Issy-Les-Moulineaux, France: Nuclear Energy Agency Organisation for Economic Co-Operation and Development, OECD; 2004. Available from: https://oecd-nea.org/dbtdb/pubs/uranium.pdf [Accessed: 2018-05-15]

[73] Guillaumont NYR, Fanghänel T, Neck V, Fuger J, Palmer DA, Grenthe I, Rand MH. In: Mompean FJ, Illemassene M, Domenech-Orti C, Ben Said K, editors. Update on the Chemical Thermodynamics of Uranium, Neptunium, Plutonium, Americium, and Technetium. Issy-Les-Moulineaux, France: OECD Nuclear Energy Agency, Data Bank; 2003. Available from: https://www.oecd-nea.org/dbtdb/pubs/vol5-update-combo.pdf [Accessed: 2018–05-15]

[74] Chevalier PY, Fischer E, Cheynet B. Progress in the thermodynamic modelling of the O–U binary system. Journal of Nuclear Materials. 2002;**303**:1-28. DOI: 10.1016/S0022-3115(02)00813-9

[75] Konings RJM, Benes O, Kovacs A, Manara D, Gorokhov DS, Iorish VS, Yungman V, Shenyavskaya E, Osina E. The thermodynamic properties of the f-elements and their compounds. Part 2. The lanthanide and actinide oxides. Journal of Physical and Chemical Reference Data. 2014;**43**:013101. DOI: 10.1063/1.4825256

[76] Wanner H, Forest I, editors. Chemical Thermodynamics of Uranium. Amsterdam: Elsevier Science Publishers B.V; 1992. ISBN: 9780444893819

[77] Murphy WM, Pabalan RT. Review of Empirical Thermodynamic Data for Uranyl Silicate Minerals and Experimental Plan. Nuclear Regulatory Commission Contract NRC-02-93-005.

San Antonio, Texas: Center for Nuclear Waste Regulatory Analyses; 1995. Available from: https://www.nrc.gov/docs/ML0402/ML040210708.pdf [Accessed: 2018-05-15]

[78] Kubatko KA, Helean KB, Navrotsky A, Burns PC. Thermodynamics of uranyl minerals: Enthalpies of formation of rutherfordine, UO_2CO_3, andersonite, $Na_2CaUO_2(CO_3)_3(H_2O)_5$, and grimselite, $K_3Na(UO_2)(CO_3){\cdot}3H_2O$. American Mineralogist. 2005;**90**:1284-1290. DOI: 10.2138/am.2005.1821

[79] Gorman-Lewis D, Shvareva TY, Kubatko KA, Burns PC, Wellman DM, McNamara B, Szymanowsky JES, Navrotsky A, Fein JB. Thermodynamic properties of autunite, uranyl hydrogen phosphate, and uranyl orthophosphate from solubility and calorimetric measurements. Environmental Science and Technology. 2009;**43**:7416-7422. DOI: 10.1021/es9012933

[80] Nguyen SN, Silva RJ, Weed HC, Andrews LE. Standard Gibbs free energies of formation at the temperature 303.15 K of four uranyl silicates: Soddyite, uranophane, sodium Boltwoodite, and sodium Weeksite. Journal of Chemical Thermodynamics. 1992;**24**:259-276. DOI: 10.1016/S0021-9614(05)80155-7

[81] Prikryl JD. Uranophane dissolution and growth in $CaCl_2$–SiO_2(aq) test solutions. Geochimica et Cosmochimica Acta. 2008;**72**:4508-4520. DOI: 10.1016/j.gca.2008.06.022

[82] Shvareva TY, Mazeina L, Gorman-Lewis D, Burns PC, Szymanowski JES, Fein JB, Navrotsky A. Thermodynamic characterization of Boltwoodite and uranophane: Enthalpy of formation and aqueous solubility. Geochimica et Cosmochimica Acta. 2011; **75**:5269-5282. DOI: 10.1016/j.gca.2011.06.041

[83] Szenknect S, Mesbah A, Cordara T, Clavier N, Brau HP, Le Goffa X, Poinssot C, Ewing RC, Dacheux N. First experimental determination of the solubility constant of coffinite. Geochimica et Cosmochimica Acta. 2016;**81**:36-53. DOI: 10.1016/j.gca.2016.02.010

[84] Guo X, Szenknect S, Mesbah A, Labs S, Clavier N, Poinssot C, Ushakova SV, Curtius H, Bosbach D, Ewing RC, Burns PC, Dacheux N, Navrotsky A. Thermodynamics of formation of coffinite, $USiO_4$. Proceedings of the National Academy of Sciences of the United States of America. 2014;**112**:6551-6555

[85] Gueneau C, Baichi M, Labroche D, Chatillon CB, Sundman B. Thermodynamic assessment of the uranium–oxygen system. Journal of Nuclear Materials. 2002;**304**:161-175. DOI: 10.1016/S0022-3115(02)00878-4

[86] Loukusa H, Ikonen T, Valtavirta V, Tulkki V. Thermochemical modeling of nuclear fuel and the effects of oxygen potential buffers. Journal of Nuclear Materials. 2016;**481**:101-110. DOI: 10.1016/j.jnucmat.2016.09.014

[87] Piro MHA, Besmann TM, Simunovic S, Lewis BJ, Thompson WT. Numerical verification of equilibrium thermodynamic computations in nuclear fuel performance codes. Journal of Nuclear Materials. 2011;**414**:399-407. DOI: 10.1016/j.jnucmat.2011.05.012

[88] Piro MHA, Banfield J, Clarno KT, Thompson WT. Coupled thermochemical, isotopic evolution and heat transfer simulations in highly irradiated UO_2 nuclear fuel. Journal of Nuclear Materials. 2013;**441**:240-251

[89] Piro MHA, Banfield J, Clarno KT, Simunovic S, Besmann TM. Simulation of thermochemistry and isotopic evolution of irradiated nuclear fuel. Transactions of the American Nuclear Society. 2013;**108**:367-370. ISSN: 0003-018X

[90] Piro MHA, Leitch BW. Conjugate heat transfer simulations of advanced research reactor fuel. Nuclear Engineering and Design. 2014;**274**:30-43. DOI: 10.1016/j.nucengdes.2014.03.054

[91] Corcoran EC, Lewis BJ, Thompson WT, Mouris J, He Z. Controlled oxidation experiments of simulated irradiated UO_2 fuel in relation to thermochemical modelling. Journal of Nuclear Materials. 2011;**414**:73-82. DOI: 10.1016/j.jnucmat.2010.11.063

[92] Corcoran EC, Kaye MH, Piro MHA. An overview of thermochemical modelling of CANDU fuel and applications to the nuclear industry. Calphad. 2016;**55**:52-62. DOI: 10.1016/j.calphad.2016.04.010

[93] Besmann TM, Bartel TJ, Bertolus M, Blanc V, Bouineau V, Carlot G, et al. State-of-the-Art Report on Multi-Scale Modelling of Nuclear Fuels. Nuclear Science NEA/NSC/R/(2015)5. Nuclear Energy Agency—Energy Agency Organisation for Economic Co-operation and Development; 2015. Available from: https://www.oecd-nea.org/science/docs/2015/nsc-r2015-5.pdf [Accessed: 2018-05-15]

[94] Besmann TM, McMurray JW, Simunovic S. Application of thermochemical modeling to assessment/evaluation of nuclear fuel behavior. Calphad. 2016;**55**:47-51. DOI: 10.1016/j.calphad.2016.04.004

[95] Smith AL, Gueneau C, Fleche JL, Chatain S, Benes O, Konings RJM. Thermodynamic assessment of the Na-O and Na-U-O systems: Margin to the safe operation of SFRs. Journal of Chemical Thermodynamics. 2017;**114**:93-115. DOI: 10.1016/j.jct.2017.04.003

[96] Weck PF, Kim E. Layered uranium(VI) hydroxides: Structural and thermodynamic properties of dehydrated schoepite α-$UO_2(OH)_2$. Dalton Transactions. 2014;**43**:17191-17199. DOI: 10.1039/C4DT02455A

[97] Weck PF, Kim E. Uncloaking the thermodynamics of the studtite to metastudtite shear-induced transformation. Journal of Physical Chemistry C. 2016;**120**:16553-16560. DOI: 10.1021/acs.jpcc.6b05967

[98] Beridze G, Kowalski PM. Benchmarking the DFT+U method for thermochemical calculations of uranium molecular compounds and solids. Journal of Physical Chemistry A. 2014;**118**:11797-11810. DOI: 10.1021/jp5101126

[99] Colmenero F, Bonales LJ, Cobos J, Timón V. Thermodynamic and mechanical properties of rutherfordine mineral based on density functional theory. Journal of Physical Chemistry C. 2017;**121**:5994-6001. DOI: 10.1021/acs.jpcc.7b00699

[100] Colmenero F, Bonales LJ, Cobos J, Timón V. Density functional theory study of the thermodynamic and Raman vibrational properties of γ-UO_3 polymorph. Journal of Physical Chemistry C. 2017;**121**:14507-14516. DOI: 10.1021/acs.jpcc.7b04389

[101] Colmenero F. Characterization of secondary phases of spent nuclear fuel under final geological disposal conditions [PhD thesis]. Madrid: Universidad Autónoma de Madrid; 2017. 443 pp. DOI: 10.13140/RG.2.2.10526.43843

[102] Colmenero F, Fernández AM, Cobos J, Timón V. Thermodynamic properties of uranyl-containing materials based on density functional theory. Journal of Physical Chemistry C. 2018;**122**:5254-5267. DOI: 10.1021/acs.jpcc.7b12341

[103] Colmenero F, Fernández AM, Cobos J, Timón V. Temperature-dependent Gibbs free energies of reaction of uranyl-containing materials based on density functional theory. Journal of Physical Chemistry C. 2018;**122**:5268-5279. DOI: 10.1021/acs.jpcc.7b12368

[104] Colmenero F, Fernández AM, Cobos J, Timón V. Periodic DFT study of the thermodynamic properties and stability of schoepite and metaschoepite mineral phases. ACS Earth and Space Chemistry. 2018; Under review

[105] Colmenero F, Fernández AM, Cobos J, Timón V. Becquerelite mineral phase: Crystal structure and thermodynamic and mechanic stability by using periodic DFT. RSC Advances. 2018;**8**:24599-24616. DOI: 10.1039/c8ra04678f

[106] Benedict M, Pigford TH, Levi HW. Nuclear Chemical Engineering. New York: McGraw-Hill; 1981. ISBN: 0070045313

[107] Schulz WW. Uranium Processing; in Encyclopedia Britannica. Available from: https://global.britannica.com/technology/uranium-processing [Accessed: 2018-05-15]

[108] McNamara B, Buck EC, Hanson B. Observation of studtite and metastudtite on spent fuel. MRS Online Proceedings Library. 2002;**757**:401-406. DOI: 10.1557/PROC-757-II9.7

[109] Hanson BD, McNamara B, Buck EC, Friese JI, Jenson E, Krupka K, Arey BW. Corrosion of commercial spent nuclear fuel. 1. Formation of studtite and metastudtite. Radiochimica Acta. 2005;**93**:159-168. DOI: 10.1524/ract.93.3.159.61613

[110] Forbes TZ, Horan P, Devine T, McInnis D, Burns PC. Alteration of dehydrated schoepite and soddyite to studtite, $[(UO_2)O_2(H_2O)_2](H_2O)_2$. American Mineralogist. 2011;**96**:202-206. DOI: 10.2138/am.2011.3517

[111] Weck PF, Kim E, Jové-Colón CF, Sassani DC. Structures of uranyl peroxide hydrates: A first-principles study of studtite and metastudtite. Dalton Transactions. 2012;**41**:9748-9752. DOI: 10.1039/C2DT31242E

[112] Abrefah J, Marschmann S, Jenson ED. Examination of the Surface Coatings Removed from K-East Basin Fuel Elements. Report of the U.S. Department of Energy, Pacific Northwest National Laboratory PNNL-11806. US-DOE: Richland, Washington; 1998

[113] Schmidt AJ, Delegard CH. Assessment of K Basin Sludge Volume Expansion Resulting from Uranium Corrosion during Storage. Report of the U.S. Department of Energy, Pacific Northwest National Laboratory PNNL-13786. US-DOE: Richland, Washington; 2002

[114] Delegard CH, Schmid AJ. Uranium Metal Reaction Behavior in Water, Sludge, and Grout. Report of the U.S. Department of Energy, Pacific Northwest National Laboratory PNNL-17815. US-DOE: Richland, Washington; 2008

[115] Delegard CH, Schmidt AJ, Chenault JW. Mechanical Properties of K Basin Sludge Constituents and their Surrogates. Report of the U.S. Department of Energy, Pacific Northwest National Laboratory PNNL-14947. US-DOE: Richland, Washington; 2004

[116] Cantrell KJ, Krupka KM, Deutsch WJ, Lindberg MJ. Residual waste from Hanford tanks 241-C-203 and 241-C-204. 2. Contaminant release model. Environmental Science and Technology. 2006;**40**:3755-3761. DOI: 10.1021/es0511568

[117] Burakov BE, Strykanova EE, Anderson EB. Secondary uranium minerals on the surface of Chernobyl "lava". MRS Online Proceedings. Library. 1997;**465**:1309-1311. DOI: 10.1557/PROC-465-1309

[118] Grenthe I, Drozdzynski J, Fujino T, Buck EC, Albrecht-Schmitt TE, Wolf SF. Uranium. In: Morss LR, Edelstein NM, Fuger J, editors. The Chemistry of Actinide and Transactinide Elements. Vol. I. Berlin: Springer Science and Business Media; 2006. pp. 253-638. ISBN 978-94-007-0211-0, Chapter V

[119] Plasil J. Oxidation-hydration weathering of uraninite: The current state-of knowledge. Journal of Geosciences. 2014;**59**:99-114. DOI: 10.3190/jgeosci.163

[120] Colmenero F, Bonales LJ, Cobos J, Timón V. Study of the thermal stability of studtite by in situ Raman spectroscopy and DFT calculations. Spectrochimica Acta. A. 2017;**174**:245-253. DOI: 10.1016/j.saa.2016.11.040

[121] Bonales LJ, Colmenero F, Cobos J, Timón V. Spectroscopic Raman characterization of rutherfordine: A combined DFT and experimental study. Physical Chemistry Chemical Physics. 2016;**18**:16575-16584. DOI: 10.1039/C6CP01510G

[122] Colmenero F, Bonales LJ, Cobos J, Timón V. Structural, mechanical and vibrational study of uranyl silicate mineral soddyite by DFT calculations. Journal of Solid State Chemistry. 2017;**253**:249-257. DOI: 10.1016/j.jssc.2017.06.002

[123] Colmenero F, Cobos J, Timón V. Periodic DFT study of the structure, Raman spectrum and mechanical properties of schoepite mineral. Inorganic Chemistry. 2018;**57**:4470-4481. DOI: 10.1021/acs.inorgchem.8b00150

[124] Payne MC, Teter MP, Ailan DC, Arias A, Joannopoulos JD. Iterative minimization techniques for ab initio total-energy calculations: Molecular dynamics and conjugate gradients. Review of Modern Physics. 1992;**64**:1045-1097. DOI: 10.1103/RevModPhys.64.1045

[125] Perdew JP, Burke K, Ernzerhof M. Generalized gradient approximation made simple. Physical Review Letters. 1996;**77**:3865-3868. DOI: 10.1103/PhysRevLett.77.3865

[126] Grimme S. Semiempirical GGA-type density functional constructed with a long-range dispersion correction. Journal of Computational Chemistry. 2006;**27**:1787-1799. DOI: 10.1002/jcc.20495

[127] Perdew JP, Ruzsinszky A, Csonka GI, Vydrov OA, Scuseria GE, Constantin LA, Zhou X, Burke K. Restoring the density-gradient expansion for exchange in solids and surfaces. Physical Review Letters. 2008;**100**:136406. DOI: 10.1103/PhysRevLett.100.136406

[128] Weck PF, Gordon ME, Greathouse JA, Bryan CR, Meserole SP, Rodriguez MA, Payne C, Kim E. Infrared and Raman spectroscopy of α-ZrW2O8: A comprehensive density functional perturbation theory and experimental study. Journal of Raman Spectroscopy. 2018; **48**:1373-1384. DOI: 10.1002/jrs.5396

[129] Weck PF, Kim E, Greathouse JA, Gordon ME, Bryan CR. Assessing exchange-correlation functionals for elasticity and thermodynamics of α-ZrW2O8: A density functional perturbation theory study. Chemical Physics Letters. 2018;**658**:195-199. DOI: 10.1016/j.cplett.2018.03.025

[130] Csonka GI, Perdew JP, Ruzsinszky A, Philipsen PHT, Lebègue S, Paier J, Vydrov OA, Ángyán JG. Assessing the performance of recent density functionals for bulk solids. Physical Review B. 2009;**79**:155107. DOI: 10.1103/PhysRevB.79.155107

[131] Clark SJ, Segall MD, Pickard CJ, Hasnip PJ, Probert MIJ, Refson K, Payne MC. First principles methods using CASTEP. Zeitschrift für Kristallographie-Crystalline Materials. 2005;**220**:567-570. DOI: 10.1524/zkri.220.5.567.65075

[132] Materials Studio. Available from: http://accelrys.com/products/collaborativescience/biovia-materials-studio/ [Accessed: 2018-04-30]

[133] Troullier N, Martins JL. Efficient pseudopotentials for plane-wave calculations. Physical Review B. 1991;**43**:1993-2006. DOI: 10.1103/PhysRevB.43.1993

[134] Pfrommer BG, Cote M, Louie MSG, Cohen ML. Relaxation of crystals with the quasi-Newton method. Journal of Computational Physics. 1997;**131**:233-240. DOI: 10.1006/jcph.1996.5612

[135] Monkhorst HJ, Pack JD. Special points for Brillouin-zone integration. Physical Review B. 1976;**13**:5188-5192. DOI: 10.1103/PhysRevB.13.5188

[136] Downs RT, Bartelmehs KL, Gibbs GV, Boisen MB. Interactive software for calculating and displaying X-ray or neutron powder diffractometer patterns of crystalline materials. American Mineralogist. 1993;**78**:1104-1107. ISSN: 0003-004X

[137] Baroni S, de Gironcoli S, Dal Corso SA. Phonons and related crystal properties from density-functional perturbation theory. Review of Modern Physics. 2001;**73**:515-562. DOI: 10.1103/RevModPhys.73.515

[138] Gonze X, Lee C. Dynamical matrices, born effective charges, dielectric permittivity tensors, and Interatomic force constants from density-functional perturbation theory. Physical Review B. 1997;**55**:10355-10368. DOI: 10.1103/PhysRevB.55.10355

[139] Refson K, Tulip PR, Clark SJ. Variational density-functional perturbation theory for dielectrics and lattice dynamics. Physical Review B. 2006;**73**:155114. DOI: 10.1103/PhysRevB.73.155114

[140] Lee C, Gonze X. Ab initio calculation of the thermodynamic properties and atomic temperature factors of SiO_2 α-quartz and stishovite. Physical Review B. 1995;**51**:8610-8613. DOI: 10.1103/PhysRevB.51.8610

[141] Gurevich VM, Sergeyeva EI, Gavrichev KS, Gorbunov VE, Khodakovsky IL. Low-temperature specific heat of UO_2CO_3. Russian Journal of Physical Chemistry. 1987;**61**: 856-857. ISSN: 0036-0244

[142] Cordfunke EHP, O'Hare PAG. The Chemical Thermodynamics of Actinide Elements and Compounds. Part 3. Miscellaneous Actinide Compounds. Vienna, Austria: International Atomic Energy Agency; 1978. Online Computer Library Center (OCLC) number: 150463669

[143] Cordfunke EHP, Westrum EF. The thermodynamic properties of β-UO_3 and γ-UO_3. Thermochimica Acta. 1988;**124**:285-296. DOI: 10.1016/0040-6031(88)87031-X

[144] Jones WM, Gordon J, Long EA. The heat capacities of uranium, uranium trioxide, and uranium dioxide from 15 K to 300 K. Journal of Chemical Physics. 1952;**20**:695-699. DOI: 10.1063/1.1700518

[145] Tasker IR, O'Hare PAG, Lewis BM, Johnson GK, Cordfunke EHP. Thermochemistry of uranium compounds. XVI. Calorimetric determination of the standard molar enthalpy of formation at 298.15 K, low-temperature heat capacity, and high-temperature enthalpy increments of $UO_2(OH)_2{\cdot}H_2O$ (Schoepite). Canadian Journal of Chemistry. 1988;**66**:620-625. DOI: 10.1139/v88-106

[146] Kubatko KA, Unruh D, Burns PC. Affects of hydrogen peroxide on the stability of Becquerelite. MRS Symposium Proceedings. 2006;**893**:423-428. DOI: 10.1557/PROC-0893-JJ05-19

[147] Shoesmith DW. Used Fuel and Uranium Dioxide Dissolution Studies—A Review; NWMO Technical Report 2007-03. Ontario, Canada: University of Western Ontario; 2014. Available from: https://www.researchgate.net/publication/238622817_Used_Fuel_and_Uranium_Dioxide_Dissolution_Studies_-_A_Review [Accessed: 2018-04-30]

[148] Cox JD, Wagman DD, Medvedev VA. CODATA Key Values for Thermodynamics. New York: Hemisphere Publishing Corp; 1989. ISBN: 0891167587

[149] Gorman-Lewis D, Fein JB, Burns PC, Szymanowski JES, Converse J. Solubility measurements of the uranyl oxide hydrate phases Metaschoepite, Compreignacite, Na–Compreignacite, Becquerelite, and Clarkeite. Journal of Chemical Thermodynamics. 2008;**40**:980-990. DOI: 10.1016/j.jct.2008.02.006

[150] Giammar DE, Hering JG. Influence of dissolved sodium and cesium on uranyl oxide hydrate solubility. Environmental Science and Technology. 2004;**38**:171-179. DOI: 10.1021/es0345672

[151] Sandino A, Bruno J. The solubility of $(UO_2)_3(PO_4)_2{\cdot}4H_2O(s)$ and the formation of U(VI) phosphate complexes: Their influence in uranium speciation in natural waters. Geochimica et Cosmochimica Acta. 1992;**56**:4135-4145. DOI: 10.1016/0016-7037(92)90256-I

[152] Bruno J, Sandino A. The solubility of amorphous and crystalline schoepite in neutral to alkaline aqueous solutions. MRS Symposium Proceedings. 1989;**127**:871-878. DOI: 10.1557/PROC-127-871

[153] Meinrath G, Kato Y, Kimura T, Yoshida Z. Solid-aqueous phase equilibria of uranium (VI) under ambient conditions. Radiochimica Acta. 1996;**75**:159-167. DOI: 10.1524/ract.1996.75.3.159

[154] Meinrath G, Kimura T. Behaviour of U(V1) solids under conditions of natural aquatic systems. Inorganic Chimica Acta. 1993;**204**:79-85. DOI: 10.1016/S0020-1693(00)88116-5

[155] Kramer-Schnabel U, Bischoff H, Xi RH, Marx G. Solubility products and complex formation equilibria in the systems uranyl hydroxide and uranyl carbonate at 25°C and I = 0.1 M. Radiochimica Acta. 1992;**56**:183-188. DOI: 10.1524/ract.1992.56.4.183

[156] Rai BD, Felmy AR, Hess NJ, LeGore VL, McCready DE. Thermodynamics of the U(VI)-Ca^{2+}-Cl–OH–H_2O system: Solubility product of becquerelite. Radiochimica Acta. 2002; **90**:495-503. DOI: 10.1524/ract.2002.90.9-11_2002.495

[157] Vochten R, Van Haverbeke L. Transformation of schoepite into the uranyl oxide hydrates: Becquerelite, billietite and wölsendorfite. Mineralogy and Petrology. 1990;**43**: 65-72. DOI: 10.1007/BF01164222

[158] Casas I, Bruno J, Cera E, Finch RJ, Ewing RC. Characterization and dissolution behavior of a becquerelite from Shinkolobwe. Zaire. Geochimica et Cosmochimica Acta. 1997;**61**: 3879-3884. DOI: 10.1016/S0016-7037(97)00195-6

[159] Sandino MCA, Grambow B. Solubility Equilibria in the U(VI)-Ca-K-Cl-H_2O system: Transformation of Schoepite into Becquerelite and Compreignacite. Radiochimica Acta. 1994;**66–67**:37-43. DOI: 10.1524/ract.1994.6667.special-issue.37

Chapter 6

Magnetic Ordering in Ilmenites and Corundum-Ordered Structures

Sergio Ricardo De Lazaro,
Luis Henrique Da Silveira Lacerda and
Renan Augusto Pontes Ribeiro

Additional information is available at the end of the chapter

http://dx.doi.org/10.5772/intechopen.81772

Abstract

In the last few years, the multiferroic materials have represented a very important research topic on the design of new technological devices. A better description for this kind of materials involves two or more forms of ferroic orders coupled in a single crystalline structure. The great number of studies in this field is focused on candidates that present the coupling between a magnetic order and ferroelectricity. However, these material classes are a challenging topic on first-principles calculations due to the strong correlation that arose from the unpaired electrons. Furthermore, the partial filling of d or f orbitals reduces a high localization and a strong interaction causing failures on the electronic structure prediction. The investigation of multiferroic materials aims at their application on the development of devices such as actuators, magnetic readers, sensors and data storage. Multiferroic materials are also alternatives to the production of technological applications based on spintronic. Our proposal is to show our experience in DFT simulations for magnetic states applied in oxides of ilmenites and corundum-ordered structures. Theoretical results reported for our group until now showed a good agreement with experimental results for half-metallicity, reduced band-gap, and/or ferromagnetic ordering.

Keywords: magnetism, DFT, co-linear ilmenites, oxides, semiconductor, spin, bulk, clusters, crystalline field theory degenerated

1. Introduction

Over the last few decades, the field of magnetism has assumed a remarkable importance in the field of science and technology due to the development of new technologies. Such kind of technologies are based on the capability of control not only the electrical degree of freedom of electrons (charge) but also the magnetic nature associated with the intrinsic angular momentum of a given particle so-called spin [1].

For instance, in recent years, the scientific interest on the development of devices in the field of spintronic has been intensified. The performance of these devices depends on the spin polarization of the current used for information storage; i. e. the development of spintronic devices requires an effective way of control charge-based electronic properties by magnetic field, as well as of controlling the magnetic properties by electric currents [2–4].

In this context, multiferroic materials are essentially the best candidates due to the intrinsic multifunctional features associated with such class of compounds. In a general view, the most complete definition for these materials is based on two or more ferroic ordering (i.e. ferroelectricity, ferroelasticity, ferromagnetism, ferrotoroidicity) coupled in a single crystalline phase. Nowadays, the majority of studies are strictly focused on candidates that combine a magnetic order (ferromagnetism, antiferromagnetism) with ferroelectricity and are known as magnetoelectric multiferroics. The interest on multiferroic materials was rebirthed by theoretical reports about the scarcity of such compounds, which are also responsible for explain the unusual phenomena and successfully predict new candidates. The main advances addressed to such kind of materials are novel devices such as actuators, transducers and storage devices, as well as other potential applications including multiple state memories and novel memory media [5–9].

In order to clarify the unusual phenomena's commonly observed in such complex materials, as well as to design new candidates with feasible properties, the understanding of magnetism in solid state materials plays a fundamental role due to the complexity of the so-called itinerant electrons commonly founded in transition metals.

In this chapter, we propose a theoretical point of view about the magnetism of Ilmenite and Corundum-ordered structures—widespread candidates for multiferroic and spintronic applications—focusing on the relation between crystalline and electronic structure associated with the ground state magnetic ordering attributed to the materials. The following sections are dedicated to explain the (i) magnetism in solid state materials, (ii) the magnetic ordering of Ilmenite and Corundum-ordered materials, and (iii) theoretical approaches to investigate magnetic solid state materials.

1.1. Magnetism in solid state materials: a fundamental tool for new technologies

Magnetic materials are the ideal candidates for the consolidation of spintronic technology due to the high degree of freedom of the spin in these materials. Magnetism is one of the oldest and most fundamental scientific problems not completely clear until now. Numerous theories are proposed to explain the magnetic behavior of materials, as the well-known classical formalism where all substances have small magnets that align in certain directions depending on the chemical environment. Another theory is domain-based, where a magnetic domain represents a region with uniform magnetization, separated from other domains by well-defined boundaries. This theory allowed the observation of the Curie temperature (T_C) above which the ferromagnetic domains become paramagnetic and the magnetism disappears [10, 11].

In the Quantum Theory of magnetism, the formalism is based on the quantum angular momentum of electron, so-called spin, associated with open-shell orbitals commonly founded in Transition Metals. In this way, the microscopic origin of the magnetism is the strong electron-electron interaction that arises from the chemical bond inside the crystalline structure. Moreover, the itinerant character associated with the magnetic moments can be localized or delocalized that originates dia- or paramagnetism in solid state materials. However, the collective magnetism in the perspective of ferromagnetism, antiferromagnetism and ferrimagnetism are the fundamental keys to overcome the main questions behind the solid state magnetic ordering [11] . A schematic overview about this kind of magnetic ordering is given in **Figure 1**.

A ferromagnetic (FM—**Figure 1a**) material is characterized by a spontaneous magnetization even in the absence of a magnetic field. At T = 0 K all magnetic moments are parallel ordered due to the stronger exchange interaction between them, differently of paramagnetic species that exhibit non-interacting magnetic moments. The conventional magnetic transition metal elements with ferromagnetic ordering are Mn, Fe, Ni and Co, where Mn and Fe can also show antiferromagnetic order depending upon the crystalline structure. In addition, the thermal energy eventually overcomes the exchange the electronic exchange in ferromagnets, producing a randomizing effect where the saturation magnetization goes to zero due to the disorder

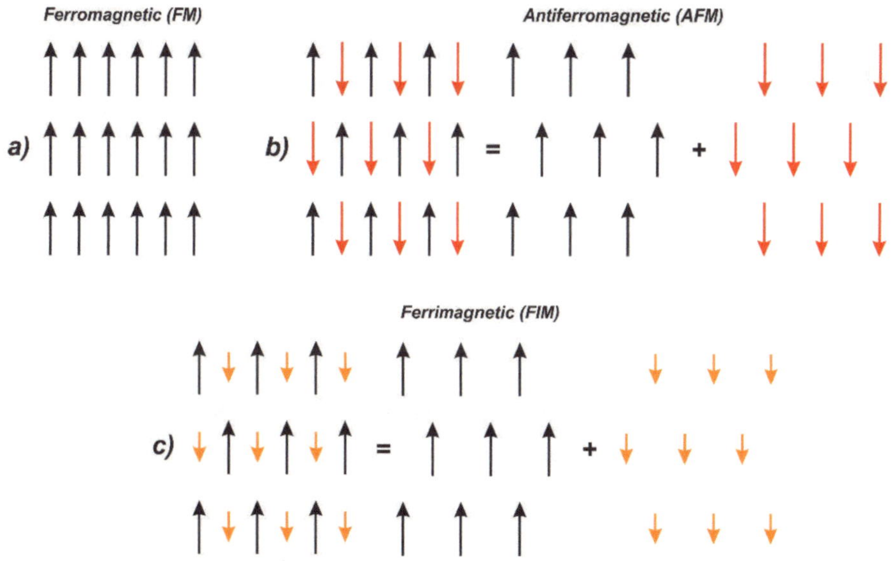

Figure 1. Different collective arrangements of magnetic moments for ordered magnetic systems: (a) ferromagnets, (b) antiferromagnets, (c) ferrimagnets.

associated with the magnetic moments. This occurs at a particular temperature called the Curie temperature (T_C) [1].

On the other hand, antiferromagnetic materials (AFM—**Figure 1b**) show an antiparallel arrangement between neighboring magnetic moments. In this case, this ordering can be attributed to the existence of negative exchange interaction between the nearest neighbors, as well as for some lattices that can be divided in two ferromagnetic sublattices, which exhibit an antiferromagnetic ordering between them. This class of compounds is the most common magnetic materials in the nature and shows good perspectives regarding the technological applications. In this class, both metals (Mn_2Au, Fe_xMn_y and others) and semiconductors/insulators (NiO, Cr_2O_3, CoO, $BiFeO_3$, MnF_2 and others) are founded [12].

Ferrimagnetism represents an intermediate position between FM and AFM orderings. In this case, the simplest picture related to solid state materials can be understood by the assumption of two magnetic sublattices with antiparallel orientation but with different magnitude of each magnetization, resulting that the total magnetization does not vanish as for the antiferromagnetic case, as represented in **Figure 1c**. Magnetite (Fe_3O_4) is a well-known ferrimagnetic material, where the spins on the tetrahedral Fe A-sublattice are antiparallel to those on the octahedral Fe B-sublattice, resulting in a net magnetic moment for the B-site [11].

Let us now briefly introduce the concept of magnetic interactions, which are responsible for the collective ordering presented in **Figure 1**. In magnetic solid state materials different types of magnetic interaction can be observed, being responsible for intriguing properties associated to the fact that magnetic moments interact between them, enabling the long range magnetic ordering. In a general point of view, two main interactions are founded in solid state materials: direct and indirect. In the first case, the electrons of neighboring magnetic atoms directly interact through an overlap between atomic wave-functions called "direct exchange". On the other hand, if the overlap of the involved wave functions is only small the direct exchange does not represent the dominating mechanism for magnetic properties because an additional atom act as a bridge between the magnetic centers, resulting in an indirect exchange interaction is responsible for magnetism. The representation of different kinds of magnetic interactions commonly founded in magnetic solid state materials is depicted in **Figure 2**.

1.1.1. Direct coupling

This type of coupling occurs when a direct overlap between the orbitals of adjacent magnetic sites is observed. The collective arrangement associated with the signal of direct coupling depends of the distance between species. At short distance the electrons tend to spend most of the time between the atoms giving rise to an antiparallel coupling due to the Pauli Exclusion Principle. However, at long distances the overlap between the orbital is reduced, inducing a minimization of potential energy for this interaction. This variation was named of Bethe-Slater curve, due to the elucidation of the behavior by these researchers. The representation of this curve is shown in **Figure 2a** [1, 11].

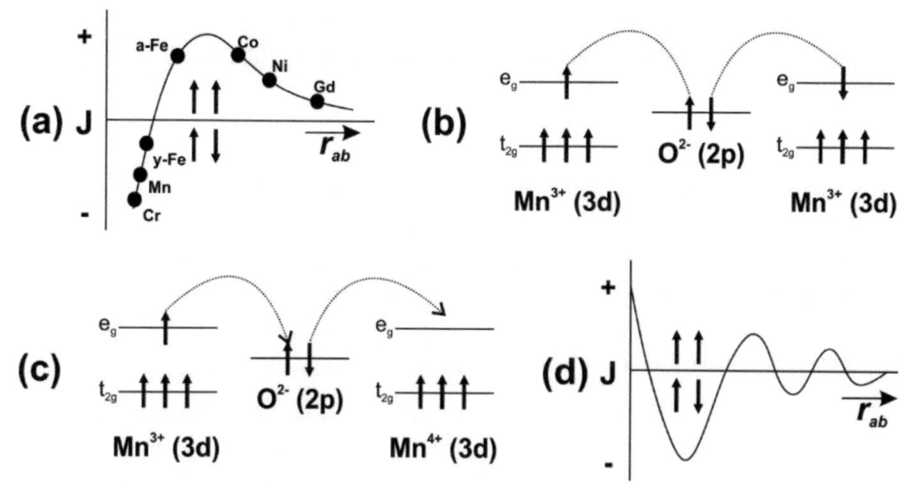

Figure 2. Representation of magnetic interactions in solid state materials. (a) Bethe-Slater curve for direct coupling. (b) Indirect coupling between Mn^{3+} mediated by oxygen atoms in MnO. (c) Oxygen-mediated indirect double-exchange coupling for cations with distinct orbital occupations. (d) RKKY coupling and its dependence with the distance between the magnetic centers.

1.1.2. Indirect coupling

In this case the electrons located in the partially occupied orbitals are separated from the neighboring electrons by an atom or non-magnetic ligand. The coupling force is dependent on the amount of energy transferred between the levels and the Coulomb repulsion between these particles. Another important feature of this model is that the oxidation numbers for the magnetic cations must be equal or have a difference equal to two, so that there is no movement between the levels. This model is present in several materials, such as the MnO shown in **Figure 2b** [1, 11].

1.1.3. Double-exchange indirect coupling

This coupling is quite similar to that previously mentioned; however the main difference lies in the movement of the electrons between the orbitals. As the oxidation numbers of the metals are different, in this case, the mismatched electrons of the A site can move to the B site, since it is possible to find unoccupied levels with the same spin orientation (**Figure 2c**) [1, 11].

1.1.4. RKKY coupling

The RKKY coupling was discovered by Ruderman, Kittel, Kasuya and Yosida (RKKY) and characterized as a long-range interaction. This type of coupling is very common in metals where the overlap between orbitals is minimal or zero, causing a polarized spin ion to induce a field on the conducting electrons of the neighboring atoms allowing a magnetic influence on a second polarized neighbor. As well as direct coupling, the orientation of neighbors is dependent on the distance between them as shown in **Figure 2d**. Another important feature of this class of magnetic couplings is the existence of effects such as Giant Magnetoresistance and Spin Tunneling that can be explained by the use of RKKY Hamiltonian [1, 11, 13–15].

1.1.5. GKA rules

The Goodenough-Kanamori-Anderson (GKA) rule describe the interatomic spin-spin interactions between two atoms considering the existence of a virtual electron transfers between them (superexchange) and/or between a shared anion (so-called "bridge") and the two atoms. As consequence, in this rule atoms with orthogonal orbitals do not overlap reducing the repulsion between the electrons, following the Hund's rule. Therefore, the exchange interaction between spins is positive and a ferromagnetic ordering arises.

In addition, the GKA rule predicts some interactions that depend on the occupation of interacting orbitals. In the first case, an antiferromagnetic interaction occurs when the virtual electron transfer is between half-filled orbitals. In opposition, they are ferromagnetic once the virtual electron transfer is from a half-filled to an empty/filled orbital.

The GKA rules are important due to the introduction of bond character in the evaluation of the magnetic interactions. It is important to point out that the net spin does not change with the covalent component associated with the chemical bonds; however this component can extends the cation wave function out because the interaction between the orbitals is large, resulting in chemical dependent ferromagnetic/antiferromagnetic couplings associated the superexchange electron transfer [16].

1.1.6. Dzyaloshinskii-Moriya rule

The Dzyaloshinskii-Moriya [17–20] rule states that a low magnetic resultant is observed in antiferromagnetic material due to a long range interaction between magnetic atoms in disordered crystalline structures. This particular behavior is called weak-Ferromagnetism.

1.2. $FeBO_3$ (B = Ti, Zr, Hf, Si, Ge, Sn) materials: insights into interatomic distance and orbital overlap

Perovskites oxides, which have the general formula ABO_3, are widely studied by theoretical or experimental efforts because of the large variety of intriguing properties, such as ferroelectricity, piezoelectricity, multiferroism and others [21–23]. The interest in this kind of structure arise from the possibility of to control the existence of different properties from chemical substitution or doping on A- and B-sites [24]. For instance, a very common mineral on earth surface ($FeTiO_3$) presents the ilmenite structure can shelter a high compositional diversity of A^{2+} and B^{4+} cations that occupy alternate basal-planes along the [001] hexagonal axis of a ordered corundum structure (R3) [25]. The most investigated ilmenite materials are based on Ti atoms in B sites with different atoms in A^{2+} (A = Mn, Fe, Ni, Co) cations; however, other materials were known in this structure [26–29].

Goodenough and Stickler proved that the magnetic ordering of ilmenite materials, mainly $ATiO_3$ (A = Mn, Fe, Ni, Co), are antiferromagnetic insulators and have two different magnetic

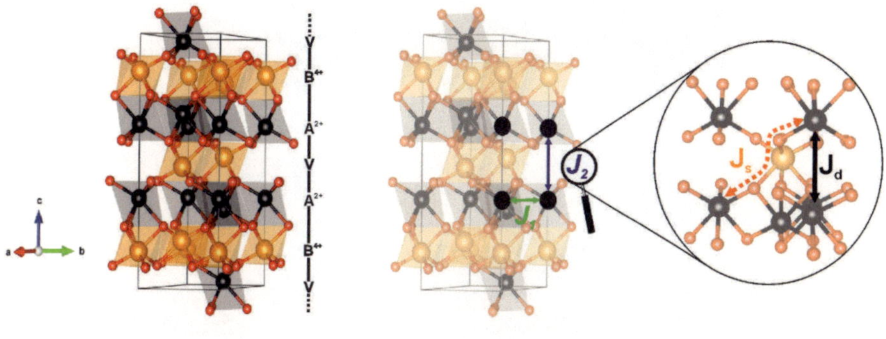

Figure 3. Ilmenite-type conventional unit cell and their exchange coupling constants. Black, orange and red balls represent A^{2+}, B^{4+} and O^{2-} ions, respectively.

couplings constants: Intralayer (J_1) and Interlayers (J_2), as shown in **Figure 1**. J_1 refers to the magnetic exchange that happens between A-O-A atoms and is dominated by the coupling of a t_{2g} orbital in one cation with an e_g orbital in other. Therefore, the signal for exchange parameter depends upon the occupancies of the interacting orbitals making $MnTiO_3$ antiferromagnetic and $FeTiO_3$, $CoTiO_3$, $NiTiO_3$ ferromagnetic for intralayer coupling (J_1). In turn, J_2 interactions are mediated by the BO_6 clusters in the intermetallic connection A-O-B-O-A, having less contribution of a direct overlap in [001] direction because of the vacancy in cationic sublattice (**Figure 3**) with opposite magnetization directions between adjacent A layers (antiferromagnetic) [25, 30].

In Fe-based ilmenite materials the long-range exchange coupling (J_2) stabilizes the antiferromagnetism; since J_2 depends of intermetallic connection Fe-O-B-O-Fe, the non-magnetic B-site replacement can control such magnetic ordering. As previously discussed, the ilmenite structure arrangement creates vacancies between adjacent Fe^{2+} layers that are separated each other by a B-site plane. So, the interlayer magnetic coupling integral can be visualized as a direct coupling between 3d orbitals of adjacent Fe^{2+} cations. Furthermore, other evidence that the non-magnetic B-site substitution affects the magnetic ordering is: the vacancy formation occurs in this cationic sublattice and the direct exchange coupling depends on the distance.

This behavior was confirmed investigating the magnetic ground-state of $FeBO_3$ (B = Ti, Zr, Hf, Si, Ge, Sn) materials by means of Density Functional Theory (DFT) calculations [31, 32]. **Figure 4** shows the relative Energy between AFM and FM structures as function of ionic radii of B^{4+} site cations.

From these results, it was observed that the FEM state is stabilized for ilmenite materials with large B-site metals (Sn, Hf, Zr); whereas, ilmenite cells contracted (Si, Ge, Ti) exhibit an AFM behavior. This result can be discussed as function of electronic repulsion between Fe^{2+} atoms 3d orbitals in different layers. In this case, we use the c-axis oriented to $3d_z{}^2$ orbital, once the cationic vacancy induces a coupling in such direction. Therefore, the B-cation volume control the distance among different Fe^{2+} layers from an angular distortion in O-B-O bonds in axial plane, which causes an increase/decrease in B-O bond distances allowing a higher/

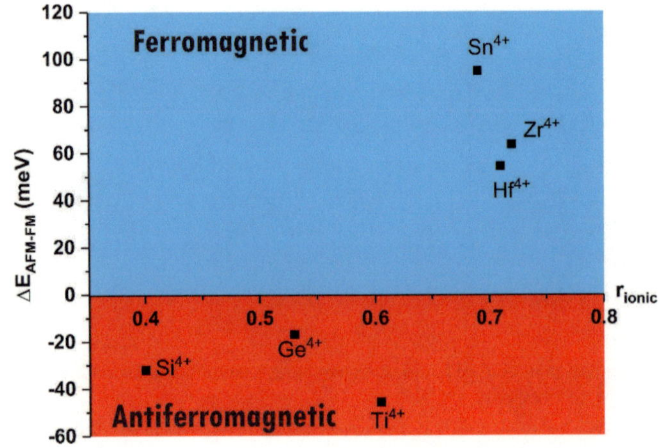

Figure 4. Energy difference (in meV) between AFM and FM configuration as function of B-site cation ionic radii (in Å).

lower interlayer distance between adjacent Fe^{2+} layers, affecting the signal of J, as represented in **Figure 5**.

In addition, the interlayer coupling has an additional degree of freedom associated with the Fe-O-B-O-Fe intermetallic connection, which is the responsible to originate a long-range coupling. This behavior can be related to the large overlap between valence orbitals from transition metals and 2p oxygen orbitals, which strengthen the intermetallic connection and, consequently, it induces a large electronic repulsion among unpaired electrons stabilizing the AFM configuration from Pauli Exclusion Principle. In case of the moving from Ti to Hf, it was noted that only $FeTiO_3$ shows a large contribution of 3d overlapped with O 2p orbital suggesting a higher overlap between these states providing strengthens in intermetallic connection that stabilizes AFM ground-state. In contrast, the smaller overlap between Zr(4d)/Hf(5d) and O(2p) orbitals allied to large Zr-O and Hf-O bond distances creates a smaller interlayer electronic repulsion responsible by FEM ordering.

Likewise, for $FeBO_3$ (B = Si, Ge, Sn) ilmenite materials is expected a FM ordering due to the absence of d valence orbitals. Nevertheless, only Sn-based ilmenite has this configuration indicating a big effect of ionic radius and interlayer distance of the FeO_6 clusters. The

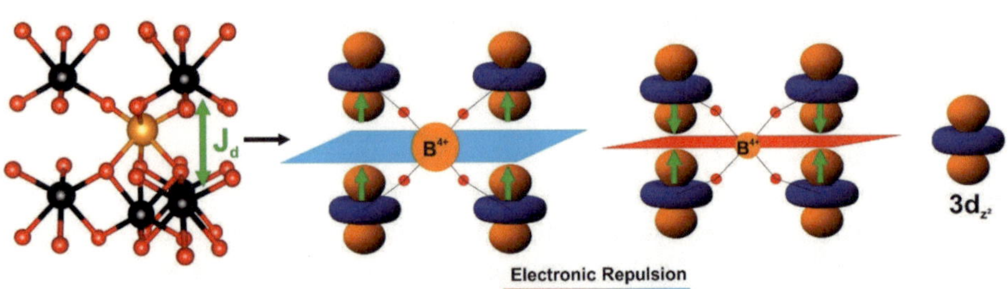

Figure 5. Representation of electronic repulsion between adjacent Fe^{2+} layer in $FeBO_3$ (B = Ti, Zr, Hf, Si, Ge, Sn) as function of ionic radius of B-site cation and its influence on magnetic ordering.

analysis of ionic radius and bond distance for $FeBO_3$ (B = Si, Ge, Sn) materials shows that the increase in ionic radius from Si^{4+} to Sn^{4+} results entails on a large distancing among Fe^{2+} layers in intermetallic connection that drastically reduce the electronic repulsion and stabilize the FM configuration. The information obtained by these theoretical results demonstrates that the control of magnetic ordering in ilmenite materials is based on a complex relation between ionic radius and valence orbitals of the B-site non-magnetic metals.

1.3. Magnetism in corundum-ordered structure ($LiNbO_3$-type structure): magnetic properties in ferroelectric $XNiO_3$ and $NiXO_3$ (X = Ti, Ge, Zr, Sn, Hf and Pb) materials

The corundum ordered structure is most commonly called $LiNbO_3$-type since this is the first material to present such crystalline phase. R3c materials are largely employed in development of memory devices, holographic data storage, electronic, electro-optical and optical devices, photo-induced devices and photocatalytic application due to a unique set of electronic, optical and ferroelectric properties usually observed for these materials. The first work on $LiNbO_3$ (LNO) structure were reported in 1949 and discuss this new structure as a ilmenite type; [33] however, in 1952 such information were refuted by Bailey [34, 35] that obtained results proving that, at room conditions, the LNO crystallizes in a R3c group instead of a R3 group characteristic of ilmenite. In particular, LNO structure has an ABO_3 general formula, lattice parameters $a = b \neq c$ and angles $\alpha = \beta = 90°$ and $\gamma = 120°$; the A and B cations are both surrounded by six O atoms forming two distorted octahedra. Some features of $LiNbO_3$-type structure are: (i) high distortion degree for octahedra within structure (**Figure 6a**); (ii) alternation between the cations A and B cation and vacancies along the c axis (**Figure 6b**); (iii) highly compacted layer composed by O atoms; (iv) the presence of intrinsic ordering vacancies [36–38]. These features are the responsible for the high ferroelectric properties characteristic of this type of structure.

The magnetism in this structure arises from unpaired electron from atoms occupying A or B sites within structure. In particular, the ground state spin ordering for materials in $LiNbO_3$-type structure is determined by collinear interaction between layers in the structure and presents only one magnetic coupling constant (J) that refers to intralayer interaction (J) that occurs between M-O-M in the structure, where M is the magnetic cation in the structure (as observed in **Figure 6c**). At general, the magnetism in Solid State materials can be successfully described by Direct or Indirect interactions, Exchange interactions and Super Exchange interactions or Goodenough-Kanamori-Anderson (GKA) Rule; all of them, considers the unit cell magnetic resultant to determine the magnetism. However, the structural disorder makes such approaches not enough to predict the magnetism in R3c structures.

The magnetic properties for $XNiO_3$ and $NiXO_3$ (X = Ti, Ge, Zr, Sn, Hf and Pb) in R3c structures were evaluated by computational methodologies based on DFT [39, 40]. The obtained results for J for such materials were presented in **Figure 7**; as observed, $TiNiO_3$, $GeNiO_3$, $ZrNiO_3$, $PbNiO_3$, $NiTiO_3$ and $NiHfO_3$ are AFM and, as predicted by DMI interaction, are weak-ferromagnetic; whereas, the other materials are FM. The evaluation of structural properties for such materials evidences a connection between structural regularity and magnetic phase

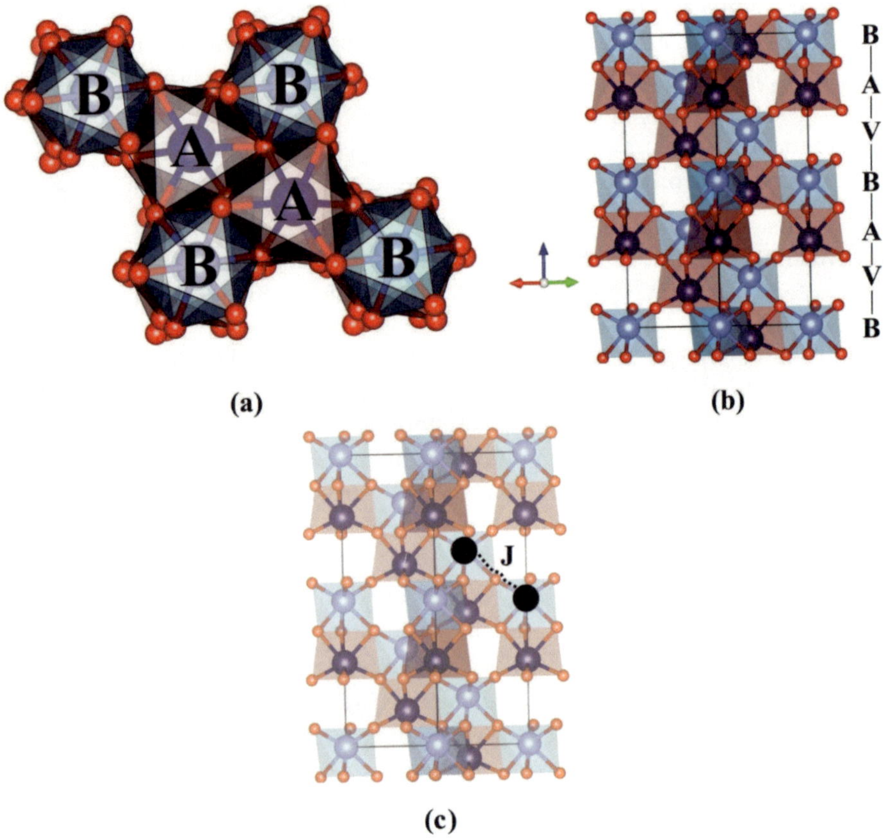

Figure 6. Crystalline structures of R3c structure showing the alternation between B, A and vacancies along z axis (a), $[AO_6]$ and $[BO_6]$ octahedra within structure (b) and magnetic coupling constant (J) for R3c structures (c).

stability, according to the relation between the distortion degree (δ) between $[AO_6]$ and $[BO_6]$, as expressed by $\delta_{[AO_6]}/\delta_{[NiO_6]}$. Thus, for materials where the value of the equation is less than 1.5, the FM phase is more stable, whereas the AFM phase is more stable for materials with disorder ratio higher than 2.

Moreover, the theoretical investigation also provides information regarding to electronic levels of magnetic cations in $XNiO_3$ and $NiXO_3$. In these materials, FM and AFM ordering are originated from e_g and t_{2g} energy levels of Ni^{2+} cations localized on $[NiO_6]$ clusters. Furthermore, projected Density of States (DOS) clarifies why magnetic ordering is changed from chemical modifications. In particular, for AFM materials there are $[NiO_6]$ clusters magnetically ordered as α and β spins (**Figure 8**); structurally, these clusters are distributed by adjacent layers being each one oriented as one spin channel. Henceforth, α and β spins localized on e_g energy levels are responsible to stabilize the AFM state.

However, how to understand the origin of the FM ordering in these materials? The existence of the FM ordering in materials is dependent on the spin orientation being in the same direction on all magnetic clusters. Thus, it is necessary to discuss the electronic configuration

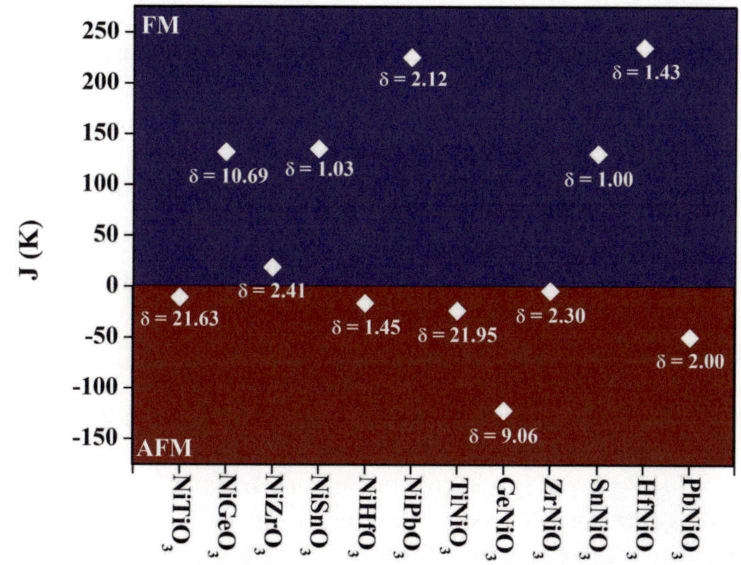

Figure 7. Magnetic coupling constant (J) for $XNiO_3$ and $NiXO_3$ (X = Ti, Ge, Zr, Sn, Hf, Pb) materials. The red region refers to the AFM materials, while the blue range represents the FM systems.

of the e_g and t_{2g} energy levels of Ni^{2+}. The projected DOS prove that in FM materials were observed a displacement on degenerated t_{2g} energy levels that creates two new groups of non-degenerated t_{2g} energy levels (**Figure 8**). In particular, the lowest t_{2g} energy levels and highest t_{2g} energy levels are occupied by α and β spins, respectively. For all $NiXO_3$ and $XNiO_3$ materials the $[NiO_6]$ clusters have the same electronic configuration, i.e., five α spins and three β spins giving rise to FM state, where all magnetic clusters are clearly at same orientation.

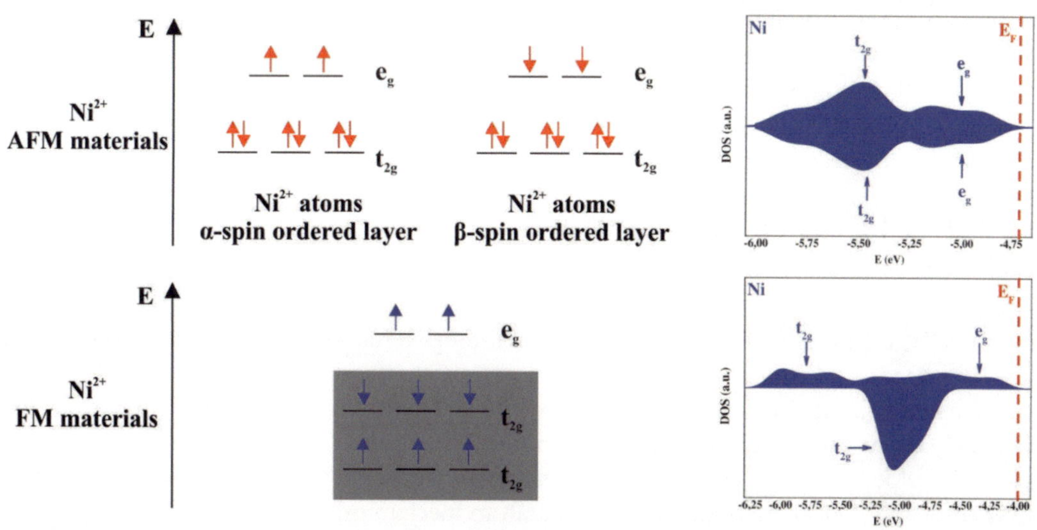

Figure 8. DOS projected results for e_g and t_{2g} energy levels localized on Ni^{2+} in AFM and FM materials. The gray region represents the broken on degeneration of t_{2g} energy levels.

Moreover, magnetic LNO-type materials have attracted the interest of materials scientist around the world in the last years as promising alternatives as smart and functional materials, mainly as multiferroic (MF) materials. The multiferroism consists on the coupling between magnetic ordering and some ferroic property in the same crystalline phase, but the main form of multiferroism is called of magnetoelectric (ME) coupling and are obtained by coupling between the magnetic ordering and ferroelectric properties. Consequently, electric polarization can be induced by a magnetic field or vice-versa.

The ME coupling was first reported in the first years of 1960s decade; thus, the study of this effect keep stable until that, in 2003, a strong coupling between ferroelectric and magnetic properties was observed for $TbMnO_3$ and $TbMn_2O_5$ materials as well as a high ferroelectricity for $BiFeO_3$ (BFO) films. Thenceforth, the study of MF materials increases drastically so that a fast search on *Web of Science* indicates at least 970 manuscripts focused in this class of material only in 2017. The investigation of multiferroic materials aims its application at development of several devices such as actuators, magnetic readers, sensors, tunneling and data storage devices. Moreover, MF materials are also potential alternatives for spintronics. The main representative of this class is the BFO; this material exhibits antiferromagnetic ordering as well as Magnetoelectric coupling at room conditions allied to a unique set of electronic, optical, magnetic and ferroelectric properties. In addition, other multiferroic materials were investigated by means of theoretical and experimental efforts, such as $YMnO_3$, $BiMnO_3$, $PbNiO_3$ (PNO), $FeTiO_3$; $PbVO_3$, $TbMnO_3$, $TbMn_2O_5$, Ca_3CoMnO_6, $LuFe_2O_4$, $BaNiF_4$ and others [41].

Even that many multiferroic materials are currently known, the development of MF material is this kind of material is delayed by crystalline structure and low Curie temperature (T_C); the first due to the fact that only 13 symmetry groups are able to exhibit multiferroic coupling, while the last makes its application impossible in technological purposes [42]. In a very restrict group of crystalline structures and, among then, are observed structures with lower and higher symmetry, such as P2/c, P-1, perovskite, ilmenite (R3) and $LiNbO_3$-type structures (R3c). It is important to highlight that the R3c structure is the most common structure observed in known multiferroic materials, such as BFO, PNO, $YMnO_3$, $BiMnO_3$ and $TbMnO_3$.

2. Theoretical approaches to investigate magnetic solid state materials

Historically, the investigation of new materials aiming technological applications was a hard task that needed a long time of study. In this period, the theoretical-computational method was seen only as tool to study of materials already discovered. Front of the fact that the technological advance is extremely dependent on development of new materials, theoretical methods helped materials chemists and physicist on development materials at higher speed. Ever since, theoretical methods based on quantum mechanical simulations are an important tool to evaluate material properties, mainly at the molecular level [43].

Nowadays, innumerous theoretical methods are available; however, in the last 30 years the Density Functional Theory (DFT) changes the world offering the best relation between the results precision and study time [44]. Another factor responsible for wide use of this theory is the high versatility, once presents good results to investigation of magnetic materials, semiconductors, proteins, organic compounds and others [45].

2.1. Density Functional Theory

Several computational approaches can be applied to evaluation of materials properties; for instance: Molecular Dynamics (MD), *ab initio* methods and Semi-Empirical Methods. The MD analyzes the system properties based on the behavior of ball-and-springs models under application of an Force external field to atoms representation, while *ab initio* and Semi-empirical methodologies uses different approaches to solve the Schrödinger Equation and to obtain the system Wave-function (Ψ). In turn, DFT assumes that the system Total Energy is a unique functional of electronic density; this methodology can be simplified in two postulates: [46].

i. The ground state properties for a system can be exactly and completely determined by the Density Functional (ρ), which is only dependent of three variables that determines the position (x, y and z).

ii. Any try function for electronic density will have energy greater or equal than the ground state energy for a real system.

However, the analytical function for electronic density is not known; thus, the electronic density is obtained by Hartree-Fock Equations (HF) for achievement of ρ by a Self Consistent Field (SCF) Method. Although this similarity was observed, the DFT shows a highest precision and computational cost in relation to HF simulations due to smaller number of variable. The HF and Semi-Empirical methods employ a number of variables in the 4n order, where n refers to the number of electrons in the system; whereas, the DFT is dependent on three variables [43, 44].

The DFT formulism was proposed by Kohn and Shan (KS) in 1965 and consists in two equations applied on two different systems; such systems consider that there are no or there are interactions between electrons, respectively. The results for both systems show a significant difference of energy between then, and, aiming to correct this difference, the Exchange-correlation Term (E_{XC}) was inserted in DFT proposed. The E_{XC} is the sum of kinetic and potential energy difference between both systems. The physical meaning of E_{XC} is the interaction between electrons in the investigated system and refers to 1% of system Total Energy; due to this, DFT describes 100% of system Total Energy.

The description of E_{XC} terms changes according to employed exchange-correlation functional and, hence, the choice of a functional has a giant effect on materials properties evaluation, as offering better results as offering a reduction in computational cost. Among the E_{XC} functional stands out local (LDA and LSDA), non-local (pure GGA, PBE) and hybrid (PBE0, B3PW91, B3LYP, HSE06) types; the last represents the most powerful alternative to predict materials project.

2.2. Magnetic materials modeling based on DFT

The DFT is widely used on materials modeling since offers an excellent relation time x efficiency, as denoted by results precision on prevision of materials properties in a short time. For instance, a DFT methodology provides thermodynamic properties for bulk or surfaces of materials directly or using some additional approach as Berry Phase [47, 48] or Bulk Modulus [49, 50]. In turn, in Solid State materials the magnetic interactions are usually described by the Heisenberg Hamiltonian; however, in computer simulations this approach cannot be used because later determinant is not an eigenstate of such Hamiltonian. Thus, it is necessary the application of the Ising model which possibilities the evaluation of J magnetic coupling constant and ground-state magnetic ordering for simple and complex magnetic structure since using the energy difference between AFM and FM magnetic states since the unpaired electrons are well defined in the model Eq. (1). The Ising Model is described by Eq. (2), where N refers to number of magnetic cations in unit cell, Z is the number of magnetic neighbors and Sx represents the spin charge for each magnetic atom in unit cell. Both Eqs. (1) and (2) are combined in Eq. (3). The application of Ising Model was performed after the structural optimization of magnetic materials and requires that the spin orientation of unpaired electrons is well defined in the model. The Ising Model is described by Eqs. (1) and (2), where N refers to number of magnetic cations in unit cell, Z is the number of magnetic neighbors and S_x represents the spin charge for each magnetic atom in unit cell. The application of Ising Model was performed after the structural optimization of magnetic materials and requires that the spin orientation of unpaired electrons is well defined in the model. Plus, in materials with more than one magnetic and not reproductive site, the Ising model should be applied to each magnetic ordering and thus, applied to ground-state model.

$$\Delta E_T = E_{T_{AFM}} - E_{T_{FM}} \tag{1}$$

$$\hat{H}_{Ising} = -N.Z.S_1.S_2.J \tag{2}$$

$$E_{T_{AFM}} - E_{T_{FM}} = -N.Z.S_1.S_2.J \tag{3}$$

The Ising Model methodology to prevision of magnetic properties of Solid State materials was employed by Ribeiro and coauthors [51], Chartier and coauthors [52], Feng and Harrison [53] and Lacerda and de Lazaro [39] to investigate materials and different crystalline structures proving the efficiency of this approach. The high efficiency and almost direct measurement offered by Ising Model indicates it as the best theoretical approach to investigation of magnetic solid state materials.

3. Conclusion

In this chapter, the role of Density Functional Theory calculations for strongly correlated materials was summarized, focusing in the description of the ground-state magnetic ordering for solid state materials, in particular $ATiO_3$ (A = Ti, Zr, Hf, Si, Ge, Sn) ilmenite derivatives and corundum ordered $XNiO_3$ and $NiXO_3$ (X = Ti, Ge, Zr, Sn, Hf and Pb) materials. A review about the theories behind the magnetism in solid state materials was presented in order to clarify the contribution of collective magnetic ordering, local structure and exchange interactions.

In case of ilmenite $ATiO_3$ derivatives, the local structure associated with the magnetic ordering of Fe^{2+} layers was systematic investigated from theoretical calculations, resulting in chemical dependent interactions depending upon the interlayer distance (B-site cation volume) and orbital-resolved electron transfer. The summation between this analysis was helpful to predict new ferromagnetic materials such as $FeSnO_{3,}$ $FeZrO_3$ and $FeHfO_{3.}$

Additionally, for corundum ordered $XNiO_3$ and $NiXO_3$ (X = Ti, Ge, Zr, Sn, Hf and Pb) materials the theoretical results indicates that the magnetic ordering is: (i) dependent on structural distortion in $[XO_6]$ and $[NiO_6]$ clusters; (ii) arise from Ni t_{2g} and e_g not conventional energy degeneration and (iii) even antiferromagnetic materials presents a magnetic resultant in consequence of structural distortion characteristic of R3c structure. The employment of theoretical approaches based on DFT possibilities the proposal of new material to technological purpose based in the magnetism as well as provides a complete description of magnetic, structural and electronic properties in a short time period.

Despite the challenge dealing with magnetic materials, theoretical simulations, mainly DFT-based, can be used to clarify the unusual phenomena's commonly observed in such materials, as well as to predict new candidates with singular properties, improving the perspective of materials design.

Author details

Sergio Ricardo De Lazaro*, Luis Henrique Da Silveira Lacerda
and Renan Augusto Pontes Ribeiro

*Address all correspondence to: srlazaro@uepg.br

Chemical Simulation Group, Department of Chemistry State University of Ponta Grossa, Paraná, Brazil

References

[1] Singh R. Unexpected magnetism in nanomaterials. Journal of Magnetism and Magnetic Materials. 2013;**346**:58-73

[2] Inomata K, Ikeda N, Tezuka N, Goto R, Sugimoto S, Wojcik M, et al. Highly spin-polarized materials and devices for spintronics. Science and Technology of Advanced Materials. 2008;**9**(1):014101

[3] Fabian J, Matos-Abiague A, Ertler C, Stano P, Žutić I. Semiconductor spintronics. Acta Physica Slovaca Reviews and Tutorials. 2007;**57**:565

[4] Felser C, Fecher Gerhard H, Balke B. Spintronics: A challenge for materials science and solid-state chemistry. Angewandte Chemie International Edition. 2007;**46**(5):668-699

[5] Ortega N, Ashok K, Scott JF, Ram SK. Multifunctional magnetoelectric materials for device applications. Journal of Physics: Condensed Matter. 2015;**27**(50):504002-504024

[6] Vopson MM. Fundamentals of Multiferroic materials and their possible applications. Critical Reviews in Solid State and Materials Sciences. 2015;**40**(4):223-250

[7] Ederer C, Spaldin NA. Recent progress in first-principles studies of magnetoelectric multiferroics. Current Opinion in Solid State and Materials Science. 2005;**9**(3):128-139

[8] Hill NA. Density functional studies of multiferroic magnetoelectrics. Annual Review of Materials Research. 2002;**32**(1):1-37

[9] Hill NA. Why are there so few magnetic ferroelectrics? The Journal of Physical Chemistry B. 2000;**104**(29):6694-6709

[10] Allenspach R, Stampanoni M, Bischof A. Magnetic domains in thin epitaxial Co/Au(111) films. Physical Review Letters. 1990;**65**(26):3344-3347

[11] Getzlaff M. Fundamentals of Magnetism. Berlin: Springer; 2007

[12] Baltz V, Manchon A, Tsoi M, Moriyama T, Ono T, Tserkovnyak Y. Antiferromagnetic spintronics. Reviews of Modern Physics. 2018;**90**(1):015005

[13] Ruderman MA, Kittel C. Indirect exchange coupling of nuclear magnetic moments by conduction electrons. Physical Review. 1954;**96**(1):99-102

[14] Kasuya TA. Theory of metallic Ferro- and Antiferromagnetism on Zener's model. Progress of Theoretical Physics. 1956;**16**(1):45-57

[15] Yosida K. Magnetic properties of Cu-Mn alloys. Physical Review. 1957;**106**(5):893-898

[16] Goodenough JB. Magnetism and the Chemical Bond. New York: Interscience Publishers; 1963

[17] Dzyaloshinsky I. A thermodynamic theory of "weak" ferromagnetism of antiferromagnetics. Journal of Physics and Chemistry of Solids. 1958;**4**(4):241-255

[18] Moriya T. Anisotropic superexchange interaction and weak ferromagnetism. Physical Review. 1960;**120**(1):91-98

[19] Sergienko IA, Dagotto E. Role of the Dzyaloshinskii-Moriya interaction in multiferroic perovskites. Physical Review B. 2006;**73**(9):094434-094438

[20] Fennie CJ. Ferroelectrically induced weak ferromagnetism by design. Physical Review Letters. 2008;**100**(16):167203-167206

[21] Bhalla AS, Guo RY, Roy R. The perovskite structure—A review of its role in ceramic science and technology. Materials Research Innovations. 2000;**4**(1):3-26

[22] Wu H, Li L, Liang L-Z, Liang S, Zhu Y-Y, Zhu X-H. Recent progress on the structural characterizations of domain structures in ferroic and multiferroic perovskite oxides: A review. Journal of the European Ceramic Society. 2015;**35**(2):411-441

[23] Zhu X. Recent patents on perovskite ferroelectric nanostructures. Recent Patents on Nanotechnology. 2009;**3**(1):42-52

[24] King G, Woodward PM. Cation ordering in perovskites. Journal of Materials Chemistry. 2010;**20**(28):5785-5796

[25] Goodenough JB. Theory of the magnetic properties of the ilmenites $MTiO_3$. Physical Review B. 1967;**164**(2):768-778

[26] Abrahams SC. Systematic prediction of new ferroelectrics in space group R3. II. Acta Crystallographica Section B-Structural Science. 2007;**63**:257-269

[27] Gou H, Zhang J, Li Z, Wang G, Gao F, Ewing RC, et al. Energetic stability, structural transition, and thermodynamic properties of $ZnSnO_3$. Applied Physics Letters. 2011;**98**(9):98-101

[28] Okada T, Narita T, Nagai T, Yamanaka T. Comparative Raman spectroscopic study on ilmenite-type $MgSiO_3$ (akimotoite), $MgGeO_3$, and $MgTiO_3$ (geikielite) at high temperatures and high pressures. American Mineralogist. 2008;**93**(1):39-47

[29] Zhang TS, Shen YS, Zhang RF. Ilmenite structure-type beta-CDSNO, used as an ethanol sensing material. Materials Letters. 1995;**23**(1-3):69-71

[30] Kato H, Yamada M, Yamauchi H, Hiroyoshi H, Takei H, Watanabe H. Metamagnetic phase transitions in $FeTiO_3$. Journal of the Physical Society of Japan. 1982;**51**(6):1769-1777

[31] Ribeiro RAP, de Lazaro SR. Structural, electronic and elastic properties of $FeBO_3$ (B = Ti, Sn, Si, Zr) ilmenite: A density functional theory study. RSC Advances. 2014; **4**(104):59839-59846

[32] Ribeiro RAP, Camilo A, de Lazaro SR. Electronic structure and magnetism of new ilmenite compounds for spintronic devices: $FeBO_3$ (B = Ti, Hf, Zr, Si, Ge, Sn). Journal of Magnetism and Magnetic Materials. 2015;**394**:463-469

[33] Matthias BT, Remeika JP. Ferroelectricity in the ilmenite structure. Physical Review. 1949;**76**(12):1886-1887

[34] Megaw HD. The structure of afwillite, $Ca_3(SiO_3OH)_2.2H_2O$. Acta Crystallographica. 1952;**5**(4):477-491

[35] Abrahams SC, Reddy JM, Bernstein JL. Ferroelectric lithium niobate. 3. Single crystal X-ray diffraction study at 24°C. Journal of Physics and Chemistry of Solids. 1966; **27**(6):997-1012

[36] Inaguma Y, Aimi A, Shirako Y, Sakurai D, Mori D, Kojitani H, et al. High-pressure synthesis, crystal structure, and phase stability relations of a $LiNbO_3$-type polar Titanate $ZnTiO_3$ and its reinforced polarity by the second-order Jahn-Teller effect. Journal of the American Chemical Society. 2014;**136**(7):2748-2756

[37] Benedek NA, Fennie CJ. Why are there so few Perovskite ferroelectrics? The Journal of Physical Chemistry C. 2013;**117**(26):13339-13349

[38] Cabuk S. The nonlinear optical susceptibility and electro-optic tensor of ferroelectrics: First-principle study. Central European Journal of Physics. 2012;**10**(1):239-252

[39] Lacerda LHS, de Lazaro SR. Multiferroism and magnetic ordering in new $NiBO_3$ (B = Ti, Ge, Zr, Sn, Hf and Pb) materials: A DFT study. Journal of Magnetism and Magnetic Materials. 2018;**465**:412-420

[40] da Silveira Lacerda LH, de Lazaro SR. Improvement of multiferroic property and change of magnetic ordering in new ANiO3 (A = Ti, Ge, Zr, Sn, Hf and Pb). Computational Materials Science. 2018;**153**:228-234

[41] Eerenstein W, Mathur ND, Scott JF. Multiferroic and magnetoelectric materials. Nature. 2006;**442**(7104):759-765

[42] Goodenough JB, Zhou J. Varied roles of Pb in transition-metal $PbMO_3$ perovskites (M = Ti, V, Cr, Mn, Fe, Ni, Ru). Science and Technology of Advanced Materials. 2015;**16**(3):036003

[43] Hautier G, Jain A, Ong S. From the computer to the laboratory: Materials discovery and design using first-principles calculations. Journal of Materials Science. 2012; **47**(21):7317-7340

[44] Burke K. Perspective on density functional theory. The Journal of Chemical Physics. 2012;**136**(15):150901-150909

[45] Becke AD. Perspective: Fifty years of density-functional theory in chemical physics. The Journal of Chemical Physics. 2014;**140**(18):18A301

[46] Lewars EG. Computational chemistry. In: Introduction to the Theory and Applications of Molecular and Quantum Mechanics. 2nd ed. Londres: Springer; 2011. p. 665

[47] Dall'Olio S, Dovesi R, Resta R. Spontaneous polarization as a Berry phase of the Hartree-Fock wave function: The case of $KNbO_3$. Physical Review B. 1997;**56**(16):10105-10114

[48] Resta R. Macroscopic polarization in crystalline dielectrics: The geometric phase approach. Reviews of Modern Physics. 1994;**66**(3):899-915

[49] Birch F. Finite elastic strain of cubic crystals. Physical Review. 1947;**71**(11):809-824

[50] Murnaghan FD. The compressibility of media under extreme pressures. Proceedings of the National Academy of Sciences. 1944;**30**(9):244-247

[51] Ribeiro RAP, de Lazaro SR, Pianaro SA. Density Functional Theory applied to magnetic materials: Mn_3O_4 at different hybrid functionals. Journal of Magnetism and Magnetic Materials. 2015;**391**:166-171

[52] Chartier A, D'Arco P, Dovesi R, Saunders VR. Ab initio Hartree-Fock investigation of the structural, electronic, and magnetic properties of Mn_3O_4. Physical Review B. 1999;**60**(20):14042-14048

[53] Feng X, Harrison NM. Magnetic coupling constants from a hybrid density functional with 35% Hartree-Fock exchange. Physical Review B. 2004;**70**(9):092402

Chapter 7

Role of Density Functional Theory in "Ribocomputing Devices"

Ruby Srivastava

Additional information is available at the end of the chapter

http://dx.doi.org/10.5772/intechopen.80491

Abstract

Molecular computing devices composed of biological substances, such as nucleic acid and ribonucleic acid plays a key role for the logical processing of a variety of inputs and viable outputs in the cellular machinery of all living organisms. These devices are directly dependent on the advancement in DNA and RNA technology. RNA nanoparticles can be engineered into a programmable and logically acting "Ribocomputing Devices"; a breakthrough at the interface of nanotechnology and synthetic biology. It opens a new path to the synthetic biologists to design reliable synthetic biological circuits which can be useful as the electronic circuits. In this emerging field, a number of challenges persist; as how to translate a variety of nucleic acid based logic gates developed by numerous research laboratories into the realm of silicon-based computing. So in this chapter we will discuss the advances in ribonucleic acid (RNA) based computing and it's potential to serve as an alternative to revolutionize silicon-based technology by theoretical means. Also the results of the calculated parameters with computational tools using Density functional theory and the designed device circuits will be analyzed.

Keywords: genetic information, logic gates, toehold switches, energy landscapes, RNA nanoparticles

1. Introduction

RNA nanotechnology is the modern and recent field of science which applies the application of top-down or bottom-up approaches to build the artificial architectures of RNA at the nanoscale. The pioneer work on "RNA nanotechnology" was carried out by eminent Scientist Prof. Dr. P. Guo et al. [1–4] RNA nanotechnology involves the characterization of the physical, chemical and biological properties of artificial RNA scaffolds in computing devices.

These molecular computers are natural and/or artificial devices in which macromolecules, including proteins and nucleic acids mediate necessary functions. The three basic operations are similar to the computer operations as: sensing inputs, processing the inputs and generating specific outputs. The schematic representation of RNA nanoparticles with different constructed motifs are given in **Figure 1**.

Nature has given us the best computer in the form of living cell. These cells generate enormous amount of signals (inputs) using a broad range of environmental factors, example: temperature, pH, pressure, nutrients, signaling chemicals, macromolecules, etc. [5–7]. Biological systems have the ability to adapt to new information from an altered environment. They can be self-assemble and self-reproduce, which might provide some economic advantages. Recently Eminent Prof. Alexander A. Green group has designed the ribocomputing devices in which RNA molecules were used as input signals and protein as the output signal. AND, OR, and NOT logic results were obtained from the self-assembly of input and gate RNAs in the

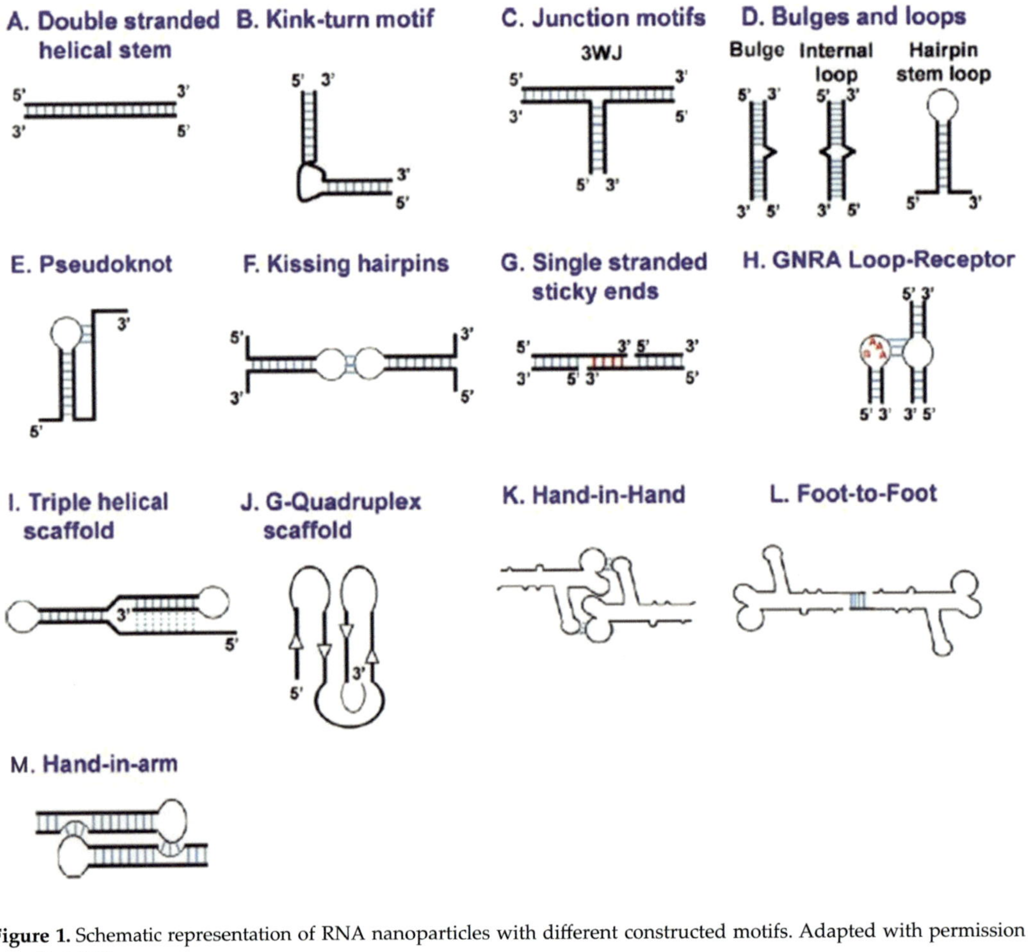

Figure 1. Schematic representation of RNA nanoparticles with different constructed motifs. Adapted with permission from Ref. [96].

device. Gate RNA was used to carry out the signal processing. These devices were operated at the post-transcriptional level and used an extended RNA transcript to co-localize all circuit sensing, computation, signal transduction, and output elements in the same self-assembled molecular complex. The advantage of these systems was the reduction in diffusion-mediated signal losses, lowered metabolic cost and improved circuit reliability. These devices utilized programmable RNA molecules, allowed effective in silico designs, composed of precisely designed synthetic RNAs networks, worked at the post-transcriptional level, minimized delays and improve the reliability of signal transduction. Further these circuits co-localized to integrate multiple circuit functions within a single transcript gate RNA [8]. The input RNAs can interact cooperatively with one another to activate the gate RNA for AND logic or they can prevent for NOT logic. Toehold switch designs, which translate an output gene only if a cognate trigger RNA is expressed in the cell, were also optimized to evaluate the AND logic gate for the ribocomputing devices [9].

Now the question arises why there is a need to replace the silicon based technology?

The capabilities of digital electronic devices have increased in lock-step with Moore's Law, which states that the number of transistors in a microchip will double approximately every 2 years [10]. This will also end up reaching a limit, as the unending quest to miniaturize transistors, which also come to a halt due to the quantum tunneling effect [11]. When the distance of a gate is scaled down about 10 nm, its electrons will jump spontaneously from source to drain, and the control over the flow of electricity will be lost. To overcome this problem, it is possible to either increase the size of microchips or fabricate them as stacking microchips. Even then it will only delay the stalemate.

Further, there are few critical challenges to make a transistor at the atomic level: (i) the accuracy of computing will be affected, because the wires in the circuit will become too close and they will affect each other; (ii) the heat generated in such a small area with too many concentrated transistors will greatly affect the functions of the transistors; (iii) the energy consumption to cool the circuit board would be too high to be a burden. Also the chip temperature will impact the circuit reliability, energy consumption and system cost.

So the next option is to use the biomolecules, including DNA, RNA and proteins as the major elements in logic gate operations. The enzymatic selectivity (processing of specific chemical function) of biomolecules give them an advantage over silicon-based computing with both specificity and usability in an intracellular environment. As most of the biological reactions controlled by specific enzymes are interconnected with other functional inputs, it is possible to fabricate DNase- and/or RNase-based informational processing units. These units can be scaled-up to fabricate artificial biocomputing networks possessing variety of logic functions.

Replacing the silicon computing devices with the molecular computers has the following other advantages: these devices are subjected to the universal physical laws, have a small mass-to-volume ratio and move in viscous media rather than in a vacuum or air. Though they cannot store momentum, kinetic energy or thermal energy for a significant period of time due to their extremely small size, and must operate isothermally. Molecular computing has an enhanced ability to provide parallel computation. These systems can self-assemble and self-reproduce, which might provide some economic advantages. Moreover, cells can be engineered to sense and respond to environmental signals, even under extreme conditions such as: high temperature, high pressure, radioactivity or toxic chemicals, and have the ability to adapt to new information for a changed environment. The ultimate goals of biocomputing are the monitoring and control of biological systems. The diagnosis of diseases, drug screening and to understand the experimental systems can be done by monitoring. It is also necessary to observe the environment to detect multiple disease indicators [11, 12], and cell based biosensors using logic gates to detect arsenic, mercury and copper ion levels [13]. Another utility is to control development, cell differentiation and re-programming, which mainly depends on gene regulatory networks [14, 15], tissue engineering and tissue regeneration [16], and to control the immune system and malign growth. Logic based biological devices are also executed to detect cancer cells (e.g. small-cell lung cancer, prostate cancer, HeLa cells), and to induce selective apoptosis of these cells [11, 17, 18].

The concept of DNA molecular computation was demonstrated by Prof. Leonard Adleman, in which the ability of synthetic DNA oligonucleotides were used to solve a seven-point Hamiltonian path problem by performing a sequence of logical operations [19]. Since then, the possibility of developing a new generation of molecular logic gates and molecular computers based on the advantages of DNA molecules has started [20–29]. The DNA molecules were chosen as they are amenable to well-regulated, programmable folding, and have the unique ability to store genetic information. The massive parallel computation power and colossal memory capacity [30, 31] were also the salient features of DNA technology, though improvement in specificity of interaction and chemical stability is still to be taken into account. DNA self-assembly properties through WC base interactions can also lead to the formation of a variety of structures (tiles) [32–34]. The studies have also been carried out on the nucleobase stacking pairs and coinage metal clusters to observe the Boolean operations [35]. However, the limitations in the application of DNA-based computation is the intolerable level of execution time in performing logic gates operation and in some circuits, it require more time to reach a stable state to achieve the final value (three-layer see-saw circuits). Another challenge is the limitation of functional proteins required for catalyzing the reactions. So the field of RNA technology has been explored. RNA can be served as the polymeric material to build varieties of structures which included nanoparticles, polygons, bundles, membranes, micro sponges and arrays. The different modifications for stable RNA nanoparticles are shown in **Figure 2**. The vital significance of RNA nanotechnology relies on application of RNA biopolymers at the atomic level. In RNA, the base pairing of nucleotides can be canonical as well as non-canonical and the tertiary interactions of RNA mediate multi-way junctions, bulges and internal hairpin loops. Apart from that RNA is thermodynamically more stable due to relatively low in energy. Now RNA is considered as a subdivision of nucleic acid nanotechnology due to its diverse functions.

The reason for taking the RNA is twofold: RNA (i) possesses the structural properties of DNA and (ii) mimics the functional properties of proteins [36–40]. Interestingly the device design and the role of DFT used for DNA technology can implement to RNA nanoparticles also due to some similar structural properties. So here we will discuss the role of DFT in RNA technology based on the previous studies being carried out on the DNA technology. RNA nanoparticles also include the following properties based on these factors: introducing

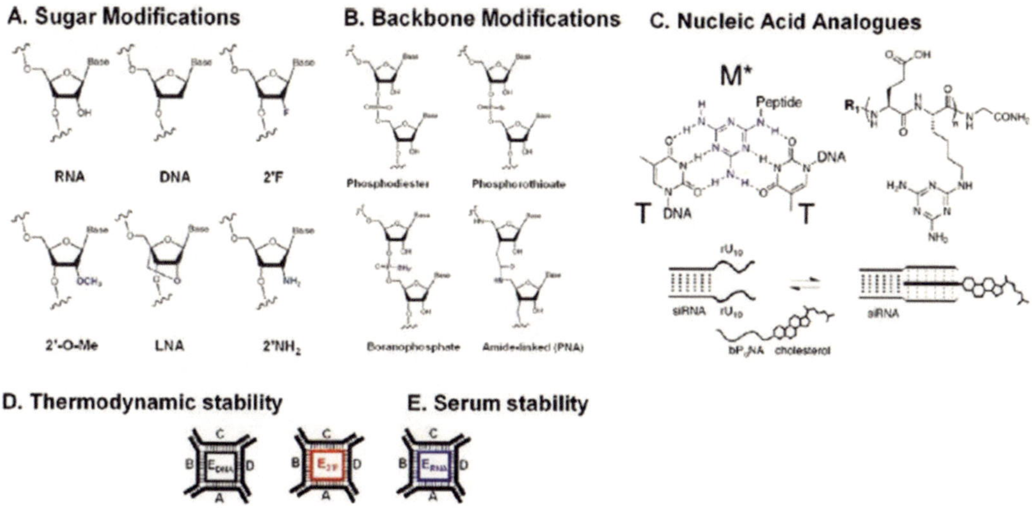

Figure 2. Different modifications for stable RNA nanoparticles (A) sugar modifications (B) backbone modifications (C) nucleic acid analogs (D) thermodynamic and (E) serum stability. Adapted with permission from Ref. [96].

chemical modification into nucleotides without significant alteration of the RNA property in folding and self-assembly; tuning the immunogenic properties of synthetic RNA constructs for in vivo applications; role of 2D, 3D, 4D structure and intermolecular interaction of RNA molecules; developing methods to control shape, size, and stoichiometry of RNA nanoparticles; regulation and processing functions of RNA in cells; cost in RNA production by biochemical synthesis; and safety of using RNA due to its therapeutic modality for cancer and other diseases without affecting the other organs.

RNA molecules have main role of passing information from genome to proteome in all living creatures. The discovery of non-coding RNAs revealed that RNA performs more versatile functions, including gene expression and regulation. The pictorial representation of beacons and resistive biomemory in RNA nanotechnology applications is given in **Figure 3**. The RNA molecules are involved in computational algorithmic processes, including RNA editing (RNA sequence alternation) [36] and RNA-based regulatory networks [35, 36]. In previous studies carried out on RNA computational systems, the inputs were small RNA elements or motifs, and the output was the mRNA [35–37].

There are various classes of RNA functional molecules, such as ribozymes, RNA aptamers, riboswitches, miRNA and siRNA, orthogonal ribosomes [38, –39] that enable the fabrication of RNA-based nanoparticles to advance the modern nanotechnology [40–42]. The recent advances are endonuclease CRISPR (Clustered regularly interspaced short palindromic repeats) associated Cas9 [43] and the bacterial CRISPR pathway [44], which enables predictable programming of gene expression. Recent efforts in genome mining and programmable RNA based switches have increased the number of parts which further enable predictable design of deeply layered logic circuits. These biological circuits required digital-like characteristics as Boolean logic gates with all or nothing responses.

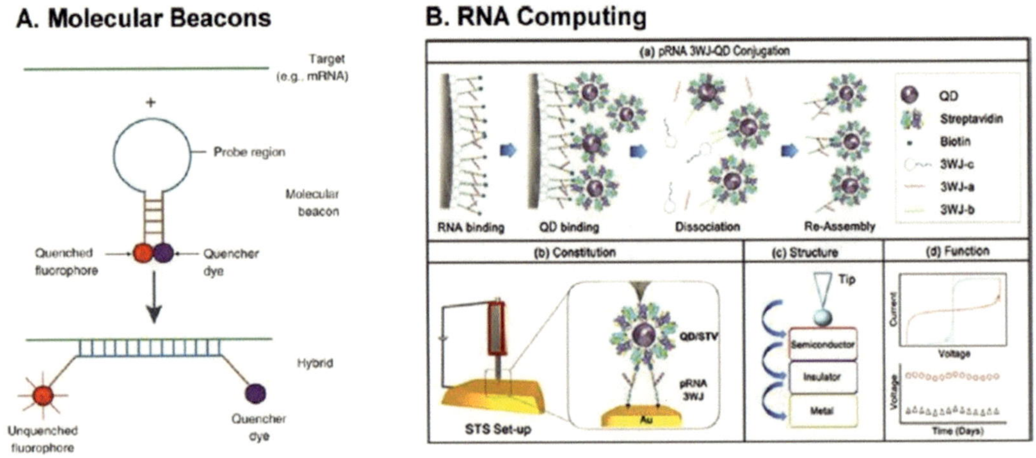

Figure 3. RNA nanotechnology applications in beacons (A) and resistive biomemory (B). Adapted with permission from Ref. [96].

2. Application of RNA nanotechnology in computation (in vivo)

RNA nanotechnology emphasizes its critical role in the process of catalysis in cell signaling and sensing functions. RNAs use the concentrations of specific chemical species as signals to implement the functions of logic to build up the network of multiple layers Boolean networks. Then these molecular computers were able to operate and communicate directly in a biological environment [45, 46] or work inside a living cell [47]. A plasmid with a specifically encoded DNA was inserted and external program was used to control the computation inside the host cell. The mRNA was used to encode a fluorescent protein and the target sequences for small interfering RNA (siRNA) with control level of fluorescence exhibited by the cell. Another in vivo computation programming was observed by a response to both endogenous and exogenous molecular signals. In this work, combination of ribozymes and RNA aptamers were used [48, 49].

The concept to utilize the cleaving ability of ribozyme to a specific molecule with the computational work was carried out using the concentrations of two proteins as input and output, implemented by studying the mechanism of ribozyme-aptamer molecules using yeast. In another study it was shown that the individual mammalian cells were capable of executing the basic arithmetic functions in a robust manner. A memory module (toggle switch) was designed [50] and a scalable factor (transcriptional/post transcriptional) was used to construct a synthetic regulatory circuit with a HeLa cancer cell 'classifier' to sense the expression levels of a customizable set of endogenous microRNAs to trigger a cellular response that matches the predetermined profile of interest [51]. In gene circuit engineering, RNA mediated regulation employs three major mechanisms, cis-acting RNA structure related modulation of translation, catalytic RNA or ribozyme-mediated cleavage of target transcripts and trans acting antisense small RNAs-mediated regulation of translation [52]. The experimental design

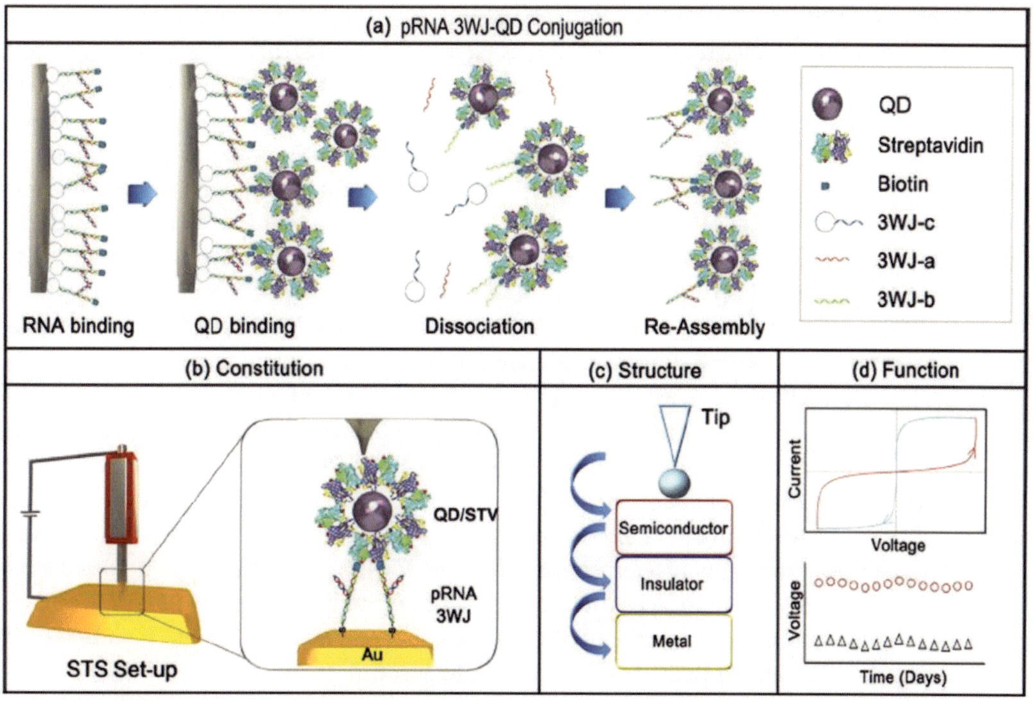

Figure 4. Schematic diagram of the experimental design method for the QD/STVQD/STV/Bio-3WJ chimera nanoparticle (a) site-specification, (b) constitution, (c) expected structure, and (d) function. Adapted with permission from Ref. [96].

method for the QD/STVQD/STV/Bio-3WJ chimera nanoparticle is given in **Figure 4**. RNA bioelectronics can lead to bio computer devices; as information storage devices, logic gates, field effect transistors and computation systems. An autonomous bio molecular system has been suggested to logically regulate the gene system. The system has been used for prostate cancer state detection. The basic diagnostic rule is that if the specific gene related to prostate cancer is over expressed, then ssDNA bind to their mRNA and exhibits the protein synthesis. The logical output is the release of ssDNA with modulation of gene expression, defined by the two states "positive and negative" [12]. An RNAi logical evaluator has also been used to perform Boolean operations for human kidney cells [47] and intracellular device in single mammalian cell.

3. Role of DFT in RNA nanotechnology

The RNA junction database [53], NanoTiler [54], RNA2D3D algorithms [55], RNA dynamics [56, 57] are used to build RNA nanoparticles that incorporate individual RNA motifs to defined user specifications. The energy landscape perspective of bio molecular dynamics provides a quantitative framework to consistently integrate theoretical concepts and experimental observations. Recognizing the potential of this approach, there has been decades of investigation into the character of protein folding landscapes and now the recent developments occur in the field of bionics. At theoretical level, the objective is to construct appropriate potential energy functions in order to explore the thermodynamic and kinetic behavior of the system. The discrete path sampling (DPS) procedure [58, 59] was employed to explore the energy landscape of the DNA G-quadruplex by HiRE-RNA force field. In order to fold proteins on physiological timescales, the energy landscape associated with folding should possess only small traps, relative to the energetic gap between the folded and unfolded ensembles. So, these landscapes should be minimally frustrated. These software's can demonstrate the biophysical and biochemical activities and also able to answer the complex analytical questions. The designed hybrid systems as (RNA–RNA, RNA–DNA, RNA–PS DNA etc.) are optimized and the reaction pathways, triggering and charge transfer mechanism can be carried out by DFT and TDDFT methods. NUPACK program [60] is used to calculate the thermodynamic analysis of inputs. Generally, three different theories can be used to analyze the results: density functional theory (DFT), and moving on to many-electron GW [61] methods as well as GW-inspired DFT + Σ [62] calculations. Further complex band structure (CBS) calculations are used to estimate the tunneling decay constant Σ, and Landauer-Buttiker transport calculations are used to compute the transmission spectra directly. Various effects such as device contact geometry, metal/molecule interfaces, molecular anchoring and side groups, inelastic effects, molecular conformation, and stochastic fluctuations are responsible for the results. So the parameters related to these effects are also taken into account. It is anticipated that the lowest conductance histogram peak correspond to single molecular junctions, while higher conductance peaks are attributed to junctions having multiple molecules linking the gold electrodes.

Electron transport mechanisms can be categorized as: coherent tunneling and incoherent hopping [63]. Along with these parameters, thermal transport and thermoelectric properties are also important [64, 65]. The transport properties estimated by GW formalism are calculated in previous studies [66, 67]. The transmission is based on two parameters Seebeck coefficient and thermo-electric figure of merit; and is very sensitive to energy and k-resolution of the calculations. The DFT + Σ method is considered to be more appropriate method to calculate the conductance, but is only applicable to system with small polarizability of molecule and broadened resonance [68, 69]. Another promising method time dependent density functional theory (TDDFT) is used to measure the wavelengths (absorption and emission) for fluorescence, and time dependent current-density functional theory (TDCDFT) for the measurement of the current–voltage characteristics. The conductance of systems is analyzed with the help of dynamical exchange−correlation correction [70–72]. The software tools TranSIESTA [73] and SMEAGOL [74] are used to calculate the electron transmission spectrum using the non-equilibrium Green's function (NEGF) method and SCARLET [75, 76] for the scattering-state method.

The complex band structure calculations are used to investigate tunneling in DNA strands [77]. Further studies were carried out for pyridine-gold junctions, which show less variation in conductance values [78–80]. These studies can provide us useful insights for the RNA-Au junctions also.

Single-molecular junctions are also the potential candidates for thermoelectric applications due to large phonon mode mismatch at the metal molecular interface and tunable electronic conductance properties [81–83]. The thermopower S, a derivative of the electron transmission at the Fermi level E_F, and proportional to the logarithmic F is another important transport property which measures the charge carrier type and thermoelectric responses. The transport of electron spin in molecular wires, and the effects of solvent properties can also studied using DFT [84–86] and DFT + Σ [87] calculations. Few calculations have also been carried out on the charge transfer in short duplexes DNA/DNA and DNA/RNA with virtually equivalent sequences [84–86]. G09 [88] and Turbomole [89] software's can been used to carried out the calculations using the implicit water solving model (COSMO) (the Conductor like screening model) [90] or PCM (polarizable continuum model) [91]. The DFT method can also be used to study the charge transfer mechanisms, delocalization nature of orbitals, base stacking, electronic coupling and conformational flexibility. Tersoff potential is preferred for carbon nanotubes-nucleotides-metal cluster interactions [92].

4. Conclusion

Scientists have demonstrated how living cells can be induced to carry out computations in the form of tiny robots or computers. Synthetic gene networks have been used to construct a wide range of biological devices, including molecular counters, oscillators, toggle switches, logic gates, cell classifiers and analog signal processors [93–95]. On a long-term scale, we believe that merging the research activities with a focus on the RNA as crucial molecular machinery for the cell will provide unprecedented insights into central molecular aspects of the RNA function and dynamics, ultimately enabling us to generate an integrated view of the molecular picture of the processes tuned by the RNA nanoparticles. Many natural limitations to the engineering of biological computers still remain. How do we eliminate or reduce noise in nucleic-acid based computing? An increase in the number of elements in RNA- and DNA-based logic operations is accompanied by a drastic increase in noise [96, 97]. The common approaches for noise reduction may include following: (i) better optimization of individual logic gates, (ii) utilization of network topology [98, 99] and (iii) for even larger networks, the introduction of new network elements to suppress the redundancy of the elements, thereby improving the signal-to-noise ratio. Also, the biochemical reaction time is also a critical aspect in working with a scalable circuit, which needs to be solved. With continued technological advances, it is now time to establish the rigorous physical and chemical understanding of RNA dynamics. Using the theoretical infrastructure developed in the context of protein folding, the field is positioned to extend the application of energy landscape techniques to the study of the large assemblies. Through close integration of computation and experimentation, it will be possible to dissect the complex interplay of structure and energetics during RNA function. We anticipate that such endeavors may ultimately reveal physical principles that govern biological dynamics. Finally we have a long way to go to understand nano-molecular computers in sufficient detail to be able to "reverse engineer" an existing molecular computer or design a relatively new one.

RS acknowledges the financial assistance by the DST WOS-A (SR/WOS-A/CS-69/2018). RS is also thankful to her mentor Dr. Shrish Tiwari, Bioinformatics Department, CSIR-Centre for Cellular and Molecular Biology, Hyderabad for the support.

Author details

Ruby Srivastava

Address all correspondence to: amitruby1@gmail.com

CSIR-Center for Cellular and Molecular Biology, Hyderabad, Telangana, India

References

[1] Shu D, Moll WD, Deng Z, Mao C, Guo P. Bottom-up assembly of RNA arrays and superstructures as potential parts in nanotechnology. Nano Letters. 2004;**4**:1717-1723. DOI: 10.1021/nl0494497

[2] Shu Y, Cinier M, Shu D, Guo P. Assembly of multifunctional φ29 pRNA nanoparticles for specific delivery of siRNA and other therapeutics to targeted cells. Methods. 2011;**54**:204-214. DOI: 10.1016/j.ymeth.2011.01.008

[3] Zhang F, Lemieux S, Wu X, St.-Arnaud S, McMurray CT, et al. Function of hexameric RNA in packaging of bacteriophage φ29 DNA in vitro. Molecular Cell. 1998;**2**:141-147. DOI: 10.1016/S1097-2765(00)80123-9

[4] Hendrix RW. Bacteriophage DNA packaging: RNA gears in a DNA transport machine (minireview). Cell. 1998;**94**:147-150. DOI: 10.1016/S0092-8674(00)81413-0

[5] Acker H. Mechanisms and meaning of cellular oxygen sensing in the organism. Respiration Physiology. 1994;**95**:1-10

[6] Daly K, Darby AC, Hall N, Wilkinson MC, Pongchaikul P, Bravo D, et al. Bacterial sensing underlies artificial sweetener-induced growth of gut lactobacillus. Environmental Microbiology. 2015;**18**(7):2159-2171. DOI: 10.1111/1462-2920.12942

[7] Green J, Paget MS. Bacterial redox sensors. Nature Reviews. Microbiology. 2004;**2**:954-966. DOI: 10.1038/nrmicro1022

[8] Green AA, Kim J, Ma D, Silver PA, Collins JJ, Yin P. Complex cellular logic computation using ribocomputing devices. 2017;**548**:117-133. DOI: 10.1038/nature23271

[9] Green AA, Silver PA, Collins JJ, Yin P. Toehold switches: De-novo designed regulators of gene expression. Cell. 2014;**159**:925-939. DOI: 10.1016/j.cell.2014.10.002

[10] Livstone MS, van Noort D, Landweber LF. Molecular computing revisited: A Moore's law? Trends in Biotechnology. 2003;**21**:98-101. DOI: 10.1016/S0167-7799(03)00005-2

[11] Keyes RW. Physical limits of silicon transistors and circuits. Reports on Progress in Physics. 2005;**68**:2701-2746. DOI: 10.1088/0034-4885/68/12/R01

[12] Benenson Y, Gil B, Ben-Dor U, Adar R, Shapiro E. An autonomous molecular computer for logical control of gene expression. Nature. 2004;**429**:423-429. DOI: 10.1038/nature02551

[13] Gil B, Kahan-Hanum M, Skirtenko N, Adar R, Shapiro E. Detection of multiple disease indicators by an autonomous biomolecular computer. Nano Letters. 2011;**11**:2989-2996. DOI: 10.1021/nl2015872

[14] Lewandoski M. Conditional control of gene expression in the mouse. Nature Reviews. Genetics. 2001;**2**:743-755. DOI: 10.1111/j.1600-6143.2006.01685.x

[15] Wang B, Barahona M, Buck M. A modular cell-based biosensor using engineered genetic logic circuits to detect and integrate multiple environmental signals. Biosensors & Bioelectronics. 2013;**40**:368-376. DOI: 10.1016/j.bios.2012.08.011

[16] Peter IS, Faure E, Davidson EH. Predictive computation of genomic logic processing functions in embryonic development. Proceedings of the National Academy of Sciences. 2012;**109**:16434-16442. DOI: 10.1016/j.cub.2008.02.061

[17] Khademhosseini A, Langer R, Borenstein J, Vacanti JP. Microscale technologies for tissue engineering and biology. Proceedings of the National Academy of Sciences of the United States of America. 2006;**103**:2480-2487. DOI: 10.1073/pnas.0507681102

[18] Morelli AE, Thomson AW. Tolerogenic dendritic cells and the quest for transplant tolerance. Nature Reviews. Immunology. 2007;**7**:610-621. DOI: https://doi.org/10.1159/000346507

[19] Adleman LM. Molecular computation of solutions to combinatorial problems. Science. 1994;**266**:1021-1024. DOI: 10.1126/science.7973651

[20] Wang K, Tang Z, Yang CJ, Kim Y, Fang X, Li W, et al. Molecular engineering of DNA: Molecular beacons. Angewandte Chemie (International Edition in English). 2009;**48**:856-870. DOI: 10.1002/anie.200800370

[21] Cornett EM, Campbell EA, Gulenay G, Peterson E, Bhaskar N, Kolpashchikov DM. Molecular logic gates for DNA analysis: Detection of rifampin resistance in *M. tuberculosis* DNA. Angewandte Chemie (International Edition in English). 2012;**51**:9075-9077. DOI: 10.1002/anie.201203708

[22] Ogawa A, Susaki Y. Multiple-input and visible-output logic gates using signal-converting DNA machines and gold nanoparticle aggregation. Organic & Biomolecular Chemistry. 2013;**11**:3272-3276. DOI: 10.1039/C3OB40313K

[23] Okamoto A, Tanaka K, Saito I. DNA logic gates. Journal of the American Chemical Society. 2004;**126**:9458-9463. DOI: 10.1021/ja047628k

[24] Penchovsky R. Engineering integrated digital circuits with allosteric ribozymes for scaling up molecular computation and diagnostics. ACS Synthetic Biology. 2012;**1**:471-482. DOI: 10.1021/sb300053s

[25] Zhang C, Yang J, Xu J. Circular DNA logic gates with strand displacement. Langmuir. 2010;**26**:1416-1419. DOI: 10.1021/la903137f

[26] Qian L, Winfree E. Scaling up digital circuit computation with DNA strand displacement cascades. Science. 2011;**332**:1196-1201. DOI: 10.1126/science.1200520

[27] Weitz M, Kim J, Kapsner K, Winfree E, Franco E, Simmel FC. Diversity in the dynamical behaviour of a compartmentalized programmable biochemical oscillator. Nature Chemistry. 2014;**6**:295-302. DOI: 10.1038/nchem.1869

[28] Winfree E. Algorithmic self-assembly of DNA: Theoretical motivations and 2D assembly experiments. Journal of Biomolecular Structure & Dynamics. 2000;**17**(1):263-270. DOI: 10.1080/07391102.2000.10506630

[29] Barish RD, Rothemund PW, Winfree E. Two computational primitives for algorithmic self-assembly: Copying and counting. Nano Letters. 2005;**5**:2586-2592. DOI: 10.1021/nl052038l

[30] Ajo-Franklin CM, Drubin DA, Eskin JA, Gee EP, Landgraf D, Phillips I, et al. Rational design of memory in eukaryotic cells. Genes & Development. 2007;**21**:2271-7226. DOI: 10.1101/gad.1586107

[31] Goldman N, Bertone P, Chen S, Dessimoz C, LeProust EM, Sipos B, et al. Towards practical, high-capacity, low-maintenance information storage in synthesized DNA. Nature. 2013;**494**:77-80. DOI: 10.1038/nature11875

[32] Aghebat Rafat A, Pirzer T, Scheible MB, Kostina A, Simmel FC. Surface-assisted large-scale ordering of DNA origami tiles. Angewandte Chemie (International Edition in English). 2014;**53**:7665-7668. DOI: 10.1002/anie.201403965

[33] Wei B, Dai M, Yin P. Complex shapes self-assembled from single-stranded DNA tiles. Nature. 2012;**485**:623-626. DOI: 10.1038/nature11075

[34] Wei B, Vhudzijena MK, Robaszewski J, Yin P. Self-assembly of complex two-dimensional shapes from single-stranded DNA tiles. Journal of Visualized Experiments. 8 May 2015;**99**:e52486. DOI: 10.3791/52486

[35] Srivastava R. Nucleobase pairs—(M_2 clusters and M_2^{2+} cluster ions (M = Ag, Au, Cu) interaction: A DFT and TDDFT Study. Submitted to Scientific Reports

[36] Guo P. The emerging field of RNA nanotechnology. Nature Nanotechnology. 2010;**5**: 833-842. DOI: 10.1038/nnano.2010.231

[37] Guo P. RNA nanotechnology: Methods for synthesis, conjugation, assembly and application of RNA nanoparticles. Methods. 2011;**54**:201-203. DOI: 10.1021/nn200989r

[38] Qiu M, Khisamutdinov E, Zhao Z, Pan C, Choi JW, Leontis NB, et al. RNA nanotechnology for computer design and in vivo computation. Philosophical Transactions of the Royal Society A: Mathematical, Physical and Engineering. 2013;**371**:20120310. DOI: 10.1098/rsta.2012.0310

[39] Grabow WW, Jaeger L. RNA self-assembly and RNA nanotechnology. Accounts of Chemical Research. 2014;**47**:1871-1880. DOI: 10.1021/ar500076k

[40] Afonin KA, Grabow WW, Walker FM, Bindewald E, Dobrovolskaia MA, Shapiro BA, et al. Design and self-assembly of siRNAfunctionalized RNA nanoparticles for use in automated nanomedicine. Nature Protocols. 2011;**6**:2022-2034. DOI: 10.1038/nprot.2011.418

[41] Afonin KA, Bindewald E, Yaghoubian AJ, Voss N, Jacovetty E, Shapiro BA, et al. In vitro assembly of cubic RNA-based scaffolds designed in silico. Nature Nanotechnology. 2010;**5**:676-682. DOI: 10.1038/nnano.2010.160

[42] Khisamutdinov EF, Bui MN, Jasinski D, Zhao Z, Cui Z, Guo P. Simple method for constructing RNA triangle, square, pentagon by tuning interior RNA 3WJ angle from 60 degrees to 90 degrees or 108 degrees. Methods in Molecular Biology. 2015;**1316**:181-193. DOI: 10.1007/978-1-4939-2730-2

[43] Jasinski D, Haque F, Binzel DW, Guo P. Advancement of the emerging field of RNA nanotechnology. ACS Nano. 2017;**11**:1142-1164. DOI: 10.1021/acsnano.6b05737

[44] Zegans ME, Wagner JC, Cady KC, Murphy DM, Hammond JH, O'Toole GA. Interaction between bacteriophage DMS3 and host CRISPR region inhibits group behaviors of *Pseudomonas aeruginosa*. Journal of Bacteriology. 2009;**191**:210-219. DOI: 10.1128/JB.00797-08

[45] De Silva AP. Molecular computation: Molecular logic gets loaded. Nature Materials. 2005;**4**:15-16. DOI: 10.1038/nmat1301

[46] Miyoshi D, Inoue M, Sugimoto N. DNA logic gates based on structural polymorphism of telomere DNA molecules responding to chemical input signals. Angewandte Chemie (International Edition in English). 2006;**45**:7716-7719. DOI: 10.1002/anie.200602404

[47] Rinaudo K, Bleris L, Maddamsetti R, Subramanian S, Weiss R, et al. A universal RNAi-based logic evaluator that operates in mammalian cells. Nature Biotechnology. 2007;**25**:795-801. DOI: 10.1038/srep00641

[48] Breaker RR. Engineered allosteric ribozymes as biosensor components. Current Opinion in Biotechnology. 2002;**13**:31-39. DOI: 10.1016/S0014-5793(03)01160-8

[49] Suess B, Weigand JE. Engineered riboswitches: Overview, problems and trends. RNA Biology. 2008;**5**:24-29. DOI: 10.4161/rna.5.1.5955

[50] Auslander S, Auslander D, Muller M, Wieland M, Fussenegger M. Programmable single-cell mammalian biocomputers. Nature. 2012;**487**:123-127. DOI: 10.1038/nature11149

[51] Lou C, Liu X, Ni M, Huang Y, Huang Q, et al. Synthesizing a novel genetic sequential logic circuit: A push-on push-off switch. Molecular Systems Biology. 2010;**6**:350. DOI: 10.1038/msb.2010.2

[52] Xie Z, Wroblewska L, Prochazka L, Weiss R, Benenson Y. Multiinput RNAi-based logic circuit for identification of specific cancer cells. Science. 2011;**333**:1307-1311. DOI: 10.1126/science.1205527

[53] Bindewald E, Grunewald C, Boyle B, O'Connor M, Shapiro BA. Computational strategies for the automated design of RNA nanoscale structures from building blocks using NanoTiler. Journal of Molecular Graphics & Modelling. 2008;**27**:299-308. DOI: 10.1016/j.jmgm.2008.05.004

[54] Bindewald E, Hayes R, Yingling YG, Kasprzak W, Shapiro BA. RNA junction: A database of RNA junctions and kissing loops for three-dimensional structural analysis and nanodesign. Nucleic Acids Research. 2008;**36**:D392-D397. DOI: 10.1093/nar/gkm842

[55] Martinez HM, Maizel JV, Shapiro BA. 2008 RNA2D3D: A program for generating, viewing, and comparing 3-dimensional models of RNA. Journal of Biomolecular Structure & Dynamics. 2008;**25**:669-683. DOI: 10.1080/07391102.2008.10531240

[56] Kasprzak W, Bindewald E, Kim TJ, Jaeger L, Shapiro BA. Use of RNA structure flexibility data in nanostructure modeling. Methods. 2010;**54**:239-250. DOI: 10.1016/j.ymeth.2010.12.010

[57] Paliy M, Melnik R, Shapiro BA. Molecular dynamics study of the RNA ring nanostructure: A phenomenon of self-stabilization. Physical Biology. 2009;**6**:046003. DOI: 10.1088/14783975/6/4/046003

[58] Wales DJ. Discrete path sampling. Molecular Physics. 2002;**100**:3285-3305. DOI: 10.1080/00268970210162691

[59] Wales DJ. Some further applications of discrete path sampling to cluster isomerization. Molecular Physics. 2004;**102**:891-908. DOI: 10.1080/00268970410001703363

[60] Dirks RM, Bois JS, Schaeffer JM, Winfree E, Pierce NA. Thermodynamic analysis of interacting nucleic acid strands. SIAM Review. 2007;**49**:65-88. DOI: 10.1137/060651100

[61] Hedin L. New method for calculating 1-particle greens function with application to electron-gas problem. Physical Review. 1965;**139**(3A):A796-A823. DOI: 10.1103/PhysRev.139.A796

[62] Quek SY et al. Length dependence of conductance in aromatic single-molecule junctions. Nano Letters. 2009;**9**(11):3949-3953. DOI: 10.1021/nl9021336

[63] Nozaki D et al. Disorder and dephasing effects on electron transport through conjugated molecular wires in molecular junctions. Physical Review B. 2012;**85**(15):155327. DOI: 10.1103/PhysRevB.85.155327

[64] Aradhya SV, Venkataraman L. Single-molecule junctions beyond electronic transport. Nature Nanotechnology. 2013;**8**(6):399-410. DOI: 10.1038/nnano.2013.91

[65] Lee W et al. Heat dissipation in atomic-scale junctions. Nature. 2013;**498**(7453):209-212. DOI: 10.1038/nature12183

[66] Strange M et al. Self-consistent GW calculations of electronic transport in thiol- and amine-linked molecular junctions. Physical Review B. 2011;**83**(11):115108. DOI: 10.1103/PhysRevB.83.115108

[67] Darancet P. et al. Ab initio GW electron-electron interaction effects in quantum transport. Physical Review B. 2007;**75**(7):075102. DOI:10.1103/PhysRevB.75.075102.

[68] Quek SY, Khoo KH. Predictive DFT-based approaches to charge and spin transport in single-molecule junctions and two-dimensional materials: Successes and challenges. Accounts of Chemical Research. 2016;**138**(24):7791-7795. DOI: 10.1021/ar4002526

[69] Rumetshofer M, Dorn G, Boeri L, Arrigoni E, Linden WV. First-principles molecular transport calculation for the benzenedithiolate molecule. New Journal of Physics. 2017;**19**:103007. DOI: 10.1088/1367-2630/aa8117

[70] Vignale G, Kohn W. Current-dependent exchange-correlation potential for dynamical linear response theory. Physical Review Letters. 1996;**77**(10):2037-2040. DOI: 10.1103/PhysRevLett.77.2037

[71] Sai N et al. Dynamical corrections to the DFT-LDA electron conductance in nanoscale systems. Physical Review Letters. 2005;**94**(18):186801. DOI: 10.1103/PhysRevLett.94.186810

[72] Kurth S, Stefanucci G. Dynamical correction to linear Kohn-Sham conductances from static density functional theory. Physical Review Letters. 2013;**111**(3):030601. DOI: 10.1103/PhysRevLett.111.030601

[73] Brandbyge M et al. Density-functional method for nonequilibrium electron transport. Physical Review B. 2002;**65**(16):165401. DOI: 10.1103/PhysRevB.65.165401

[74] Rocha AR et al. Spin and molecular electronics in atomically generated orbital landscapes. Physical Review B. 2006;**73**(8):085414. DOI: 10.1103/PhysRevB.73.085414

[75] Choi HJ, Cohen ML, Louie SG. First-principles scattering-state approach for nonlinear electrical transport in nanostructures. Physical Review B. 2007;**76**(15):155420. DOI: 10.1103/PhysRevB.76.155420

[76] Choi HJ, Ihm J. Ab initio pseudopotential method for the calculation of conductance in quantum wires. Physical Review B. 1999;**59**(3):2267-2275. DOI: 10.1103/PhysRevB.59.2267

[77] Wang H, Lewis JP, Sankey OF. Band-gap tunneling states in DNA. Physical Review Letters. 2004;**93**(1):016401. DOI: 10.1103/PhysRevLett.93.016401

[78] Venkataraman L et al. Single-molecule circuits with well-defined molecular conductance. Nano Letters. 2006;**6**(3):458-462. DOI: 10.1021/nl052373

[79] Quek SY et al. Mechanically controlled binary conductance switching of a singlemolecule junction. Nature Nanotechnology. 2009;**4**(4):230-234. DOI: 10.1038/nnano.2009.10

[80] Quek SY et al. Amine-gold linked single-molecule circuits: Experiment and theory. Nano Letters. 2007;**7**(11):3477-3482. DOI: 10.1021/nl072058i

[81] Wang RY, Segalman RA, Majumdar A. Room temperature thermal conductance of alkanedithiol self-assembled monolayers. Applied Physics Letters. 2006;**89**(17):173113. DOI: 10.1063/1.2358856

[82] Mahan GD, Sofo JO. The best thermoelectric. Proceedings of the National Academy of Sciences of the United States of America. 1996;**93**(15):7436-7439. DOI: 10.1073/pnas.93.15.7436

[83] Humphrey TE, Linke H. Reversible thermoelectric nanomaterials. Physical Review Letters. 2005;**94**(9):096601. DOI: 10.1103/PhysRevLett.94.096601

[84] Pilevarshahri R et al. Spin transport in higher n-acene molecules. Physical Review B. 2011;**84**(17):174437. DOI: 10.1103/PhysRevB.84.174437

[85] Bhattacharya S, Akande A, Sanvito S. Spin transport properties of triarylamine-based nanowires. Chemical Communications. 2014;**50**(50):6626-6629. DOI: 10.1039/C4CC01710B

[86] Liang SH et al. Organic magnetic tunnel junctions: The role of metal-molecule interface. Physical Review B. 2012;**86**(22):224419. DOI: 10.1103/PhysRevB.86.224419

[87] Gorjizadeh N, Quek SY. Interface effects on tunneling magnetoresistance in organic spintronics with flexible amine-Au links. Nanotechnology. 2013;**24**(41):415201. DOI: 10.1088/0957-4484/24/41/415201

[88] Frisch MJ et al. Wallingford CT: Gaussian, Inc.; 2009. (Software)

[89] Furche F, Ahlrichs R, Hättig C, Klopper W, Sierka M, Weigend F. Turbomole. WIREs Computational Molecular Science. 2014;**4**:91-100. DOI: 10.1002/wcms.1162

[90] Klamt A, Schüürmann G. COSMO: A new approach to dielectric screening in solvents with explicit expressions for the screening energy and its gradient. Journal of the Chemical Society, Perkin Transactions. 1993;**2**(5):799-805. DOI: 10.1039/P29930000799

[91] Tomasi J, Mennucci B, Cammi R. Quantum mechanical continuum solvation models. Chemical Reviews;**105**(8):2999-3094. DOI: 10.1021/cr9904009

[92] Srivastava R. Chapter 13. Novel Optoelectronic devices. 2017. DOI: 10.5772/67527. ISBN: 978-953-51-5219-4

[93] Jasinski D, Khisamutdinov EF, Lyubchenko YL, Guo P. Physicochemically tunable polyfunctionalized RNA Square architecture with fluorogenic and ribozymatic properties. ACS Nano. 2014;**8**:7620-7629. DOI: 10.1021/nn502160s

[94] Khisamutdinov EF, Jasinski DL, Li H, Zhang K, Chiu W, Guo P. Fabrication of RNA 3D nanoprisms for loading and protection of small RNAs and model drugs. Advanced Materials. 2016;**28**(45):10079-10087. DOI: 10.1002/adma.201603180

[95] Lee L, Yagati AK, Pi F, Sharma A, Choi JW, Guo P. Construction of RNAi quantum dot chimera for nanoscale resistive biomemory application. ACS Nano. 2015;**9**(7):6675-6682. DOI: 10.1021/acsnano.5b03269

[96] Li H, Zhang K, Pi F, Guo S, Shlyakhtenko L, Chiu W, et al. Controllable self-assembly of RNA tetrahedrons with precise shape and size for cancer targeting. Advanced Materials. 2016;**28**(34):7501-7507. DOI: 10.1002/adma.201601976

[97] Fedichkin L, Katz E, Privman V. Error correction and digitalization concepts in biochemical computing. Journal of Computational and Theoretical Nanoscience. 2008;**5**:36-43. DOI: 10.1166/jctn.2008.004

[98] Privman V, Arugula MA, Halamek J, Pita M, Katz E. Network analysis of biochemical logic for noise reduction and stability: A system of three coupled enzymatic and gates. The Journal of Physical Chemistry. B. 2009;**113**:5301-5310. DOI: 10.1021/jp810743w

[99] Privman V, Strack G, Solenov D, Pita M, Katz E. Optimization of enzymatic biochemical logic for noise reduction and scalability: How many biocomputing gates can be interconnected in a circuit? The Journal of Physical Chemistry. B. 2008;**112**:11777-11784. DOI: 10.1021/jp802673q